STARGAZING 2011

MONTH-BY-MONTH GUIDE TO THE NORTHERN NIGHT SKY

HEATHER COUPER & NIGEL HENBEST

www.philips-maps.co.uk

HEATHER COUPER and NIGEL HENBEST are inter-
nationally recognized writers and broadcasters on
astronomy, space and science. They have written more
than 30 books and over 1000 articles, and are the
founders of an independent TV production company
specializing in factual and scientific programming.

Heather is a past President of both the British
Astronomical Association and the Society for Popular
Astronomy. She is a Fellow of the Royal Astronomical
Society, a Fellow of the Institute of Physics and a former
Millennium Commissioner, for which she was awarded
the CBE in 2007. Nigel has been Astronomy Consultant
to *New Scientist* magazine, Editor of the *Journal of the
British Astronomical Association* and Media Consultant to
the Royal Greenwich Observatory.

Published in Great Britain in 2010
by Philip's,
a division of Octopus Publishing Group Limited
(www.octopusbooks.co.uk)
Endeavour House, 189 Shaftesbury Avenue,
London WC2H 8JY
An Hachette UK Company (www.hachette.co.uk)

TEXT
Heather Couper and Nigel Henbest (pages 6–53)
Robin Scagell (pages 61–64)
Philip's (pages 1–5, 54–60)

ISBN 978–1–84907–117–8

Printed in China

Title page: Veil Nebula (Peter Shah/Galaxy)

ACKNOWLEDGEMENTS
All star maps by Wil Tirion/Philip's, with extra
annotation by Philip's.
Artworks © Philip's.

**All photographs courtesy of
Galaxy Picture Library:**
Bob Purbrick 36;
Eddie Guscott *16, 20, 28*;
Ian King *44*;
Thierry Legault *64 top right*;
Optical Vision Ltd *64 top left*;
Philip Perkins *33*;
Robin Scagell *12, 24, 61, 62, 64 bottom right*;
Peter Shah *8, 40, 48, 53, 63*.

CONTENTS

The sight of diamond-bright stars sparkling against a sky of black velvet is one of life's most glorious experiences. No wonder stargazing is so popular. Learning your way around the night sky requires nothing more than patience, a reasonably clear sky and the 12 star charts included in this book.

Stargazing 2011 is a guide to the sky for every month of the year. Complete beginners will use it as an essential night-time companion, while seasoned amateur astronomers will find the updates invaluable.

THE MONTHLY CHARTS

Each pair of monthly charts shows the views of the heavens looking north and south. They are usable throughout most of Europe – between 40 and 60 degrees north. Only the brightest stars are shown (otherwise we would have had to put 3000 stars on each chart, instead of about 200). This means that we plot stars down to 3rd magnitude, with a few 4th-magnitude stars to complete distinctive patterns. We also show the ecliptic, which is the apparent path of the Sun in the sky.

USING THE STAR CHARTS

To use the charts, begin by locating the north Pole Star – Polaris – by using the stars of the Plough (see May). When you are looking at Polaris you are facing north, with west on your left and east on your right. (West and east are reversed on star charts because they show the view looking up into the sky instead of down towards the ground.) The left-hand chart then shows the view you have to the north. Most of the stars you see will be circumpolar, which means that they are visible all year. The other stars rise in the east and set in the west.

Now turn and face the opposite direction, south. This is the view that changes most during the course of the year. Leo, with its prominent 'sickle' formation, is high in the spring skies. Summer is dominated by the bright trio of Vega, Deneb and Altair. Autumn's familiar marker is the Square of Pegasus, while the winter sky is ruled over by the stars of Orion.

The charts show the sky as it appears in the late evening for each month: the exact times are noted in the caption with the chart. If you are observing in the early morning, you will find that the view is different. As a rule of thumb, if you are observing two hours later than the time suggested in the caption, then the following month's map will more accurately represent the stars on view. So, if you wish to observe at midnight in the middle of February, two hours later than the time suggested in the caption, then the stars will appear as they are on March's chart. When using a chart for the 'wrong' month, however, bear in mind that the planets and Moon will not be shown in their correct positions.

THE MOON, PLANETS AND SPECIAL EVENTS

In addition to the stars visible each month, the charts show the positions of any planets on view in the late evening. Other planets may also be visible that month, but they will not be on the chart if they have already set, or if they do not rise until early morning. Their positions are described in the text, so that you can find them if you are observing at other times.

We have also plotted the path of the Moon. Its position is marked at three-day intervals. The dates when it reaches First Quarter, Full Moon, Last Quarter and New Moon are given in the text. If there is a meteor shower in the month, we mark the position from which the meteors appear to emanate – the *radiant*. More information on observing the planets and other Solar System objects is given on pages 54–57.

Once you have identified the constellations and found the planets, you will want to know more about what's on view. Each month, we explain one object, such as a particularly interesting star or galaxy, in detail. We have also chosen a spectacular image for each month and described how it was captured. All of these pictures were taken by amateurs. We list details and dates of special events, such as meteor showers or eclipses, and give observing tips. Finally, each month we pick a topic related to what's on view, ranging from the Milky Way to double stars and space missions, and discuss it in more detail. Where possible, all relevant objects are highlighted on the maps.

FURTHER INFORMATION

The year's star charts form the heart of the book, providing material for many enjoyable observing sessions. For background information turn to pages 54–57, where diagrams help to explain, among other things, the movement of the planets and why we see eclipses.

Although there is plenty to see with the naked eye, many observers use binoculars or telescopes, and some choose to record their observations using cameras, CCDs or webcams. For a round-up of what's new in observing technology, go to pages 61–64, where equipment expert Robin Scagell shares his knowledge.

If you have already invested in binoculars or a telescope, then you can explore the deep sky – nebulae (starbirth sites), star clusters and galaxies. On pages 58–60 we list recommended deep-sky objects, constellation by constellation. Use the appropriate month's maps to see which constellations are on view, and then choose your targets. The table of 'limiting magnitude' (page 58) will help you to decide if a particular object is visible with your equipment.

Happy stargazing!

Early-risers this month will be treated to the sight of a dazzling Morning Star – the planet Venus. Those of us with a more nocturnal disposition will have to wait until December, when our neighbour-world reappears in the evening sky.

Plus – we'll have a partial eclipse of the Sun at sunrise on 4 January.

But the glory of the heavens this month belongs to the stars. **Orion**, **Taurus**, **Gemini** and **Canis Major** make up a scintillating celestial tableau, and there's no better time to start finding your way around the sky.

JANUARY'S CONSTELLATION

Crowned by **Sirius**, the brightest star in the sky, **Canis Major** is the larger of **Orion**'s two hunting dogs. He is represented as chasing **Lepus** (the Hare), a very faint constellation below Orion, but his main target is Orion's chief quarry, **Taurus** (the Bull) – take a line from Sirius through Orion's belt, and you'll spot the celestial bovine on the other side. Arabian astronomers accorded great importance to Canis Major, while the Indians regarded both cosmic dogs (**Canis Minor** lies to the left of Orion) as being 'watch-dogs of the Milky Way' – which runs between the two constellations.

To the right of Sirius is the star **Mirzam**. Its Arabic name means 'the Announcer,' because the presence of Mirzam heralded the appearance of Sirius, one of the most venerated stars in the sky. Just below Sirius is a beautiful star cluster, **M41**. This loose agglomeration of over a hundred young stars – 2500 light years away – is easily visible through binoculars, and even to the unaided eye. It's rumoured that the Greek philosopher Aristotle, in 325 BC, called it 'a cloudy spot' – the earliest description of a deep-sky object.

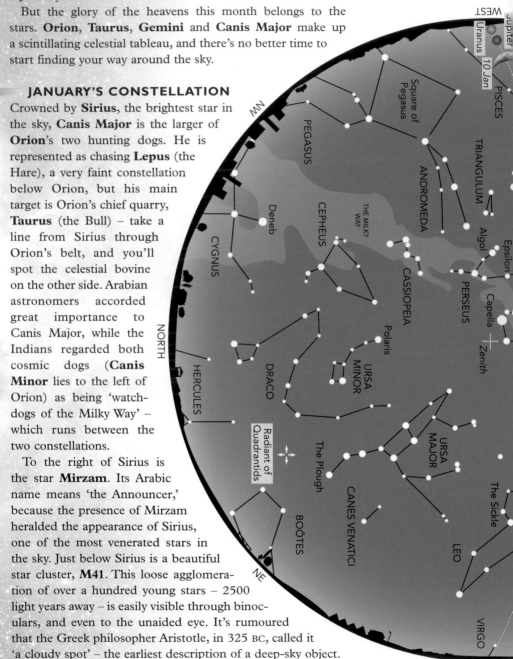

▼ *The sky at 10 pm in mid-January, with Moon positions at three-day intervals either side of Full Moon. The star positions are also correct for 11 pm at*

the beginning of January, and 9 pm at the end of the month. The planets move slightly relative to the stars during the month.

PLANETS ON VIEW

The year begins with the two brilliant planets at opposite ends of the night: Jupiter lording it over the evening sky, and Venus lighting up the dawn.

Giant planet **Jupiter** lies in Pisces: at magnitude −2.2 it far outshines any of the constellation's dim stars. At the start of January, Jupiter sets at 11 pm, but it slips below the horizon by 9.30 pm at the end of the month.

The first few days of January are also the ideal time for getting to know **Uranus**, which lies just half a degree to the upper right of Jupiter. Theoretically visible to the naked eye, binoculars will easily show you the seventh planet: at magnitude +5.9, Uranus is roughly the same brightness as Jupiter's moons.

With a telescope, you can catch **Neptune**, the most distant planet, at the beginning of January (when it sets at 8 pm), but by the end of the month it has sunk into the twilight. At magnitude +7.9, Neptune lies in Capricornus.

Saturn, shining at magnitude +1.0 in Virgo, rises at 0.40 am at the beginning of January, and by 10.45 pm at the end of the month.

For early birds, the dawn skies are the domain of brilliant **Venus**, which is at greatest elongation west on 8 January. At magnitude −4.3, the Morning Star may be bright enough to cast shadows! At the start of 2011, Venus rises at 4 am, slipping to 4.45 am by the end of January. Through a small telescope, you'll see the cloudy planet half-lit by the Sun.

WEST

Uranus
Jupiter

10 Jan

PISCES

TRIANGULUM

13 Jan

ARIES

Mira

CETUS

Algol

PERSEUS

Pleiades

ERIDANUS

16 Jan

Zeta

Aldebaran

TAURUS

Rigel

LEPUS

SW

Zenith — Capella

AURIGA

ORION

Orion Nebula

Sirius

Mirzam

CANIS MAJOR

COLUMBA

SOUTH

Castor

Betelgeuse

THE MILKY WAY

M41

Adhara

Pollux

GEMINI

19 Jan

Procyon

CANIS MINOR

PUPPIS

URSA MAJOR

CANCER

HYDRA

SE

The Sickle

Regulus

22 Jan

LEO

VIRGO

Ecliptic

EAST

	January's Object		Jupiter
	Pleiades		Uranus
	January's Picture		
	Orion Nebula		Moon
	Radiant of Quadrantids		

MOON		
Date	Time	Phase
4	9.03 am	New Moon
12	11.31 am	First Quarter
19	9.21 pm	Full Moon
26	12.57 pm	Last Quarter

During the first couple of weeks of January, you may also catch the innermost planet, **Mercury**, loitering very low in the south-east before dawn. At its greatest western elongation, on 9 January, Mercury shines at magnitude -0.2, some 40 times fainter than Venus.

Mars is too close to the Sun to be seen this month.

MOON

The crescent Moon lies near Jupiter on the evenings of 9 and 10 January. On 15 January, the Moon is near the Pleiades (Seven Sisters), and it passes Saturn on the night of 24/25 January. In the early morning of 29 January, the crescent Moon forms a striking pair with Venus, with red giant Antares to its right; the following morning, the Moon lies directly under the Morning Star.

SPECIAL EVENTS

On **3 January**, the Earth is at perihelion, its closest point to the Sun.

The maximum of the **Quadrantid** meteor shower occurs on the night of **3/4 January**. These shooting stars are tiny particles of dust shed by the old comet 2003 EH_1, burning up as they enter the Earth's atmosphere. This is an excellent year for observing the Quadrantids, as moonlight won't interfere.

On the morning of **4 January**, we're treated to an eclipse of the Sun. It's not total anywhere in the world, but people in the south-east of the UK will see a great partial eclipse – provided you've got a good view of sunrise!

The Sun is 70% eclipsed by the Moon when it pops above the horizon at 8.10 am. The Moon moves steadily away, until the eclipse ends at 9.30 am. Don't expect to see the Sun's outer atmosphere, the corona, or its flaming chromosphere – these are only visible during a total eclipse. Instead, the Sun will look as if it has had a chunk taken out of it.

DO NOT LOOK AT THE SUN DIRECTLY: PROJECT ITS IMAGE USING A PINHOLE, OR USE SPECIAL ECLIPSE GOGGLES.

> ⊙ **Viewing tip**
> Venus is a real treat this month. If you have a small telescope, though, there's no rush to point it at the Morning Star. While the sky is dark, the cloud-wreathed planet is so brilliant that it's difficult to make out anything on its disc. It's best to wait until the sky begins to brighten, and you can then see the globe of Venus appearing fainter against a pale blue sky.

JANUARY'S OBJECT

The **Pleiades** star cluster is one of the most familiar sky-sights. Though it's well known as the Seven Sisters, most people see any number of stars but seven! Nearly everyone can pick out the six brightest stars, while very keen-sighted observers can discern up to 11 stars. These are just the most luminous in a group of 500 stars, lying about 400 light years away (although there's an ongoing debate about the precise distance!). The brightest stars in the Pleiades are hot and blue, and all the stars are young – less than 80 million years old. They were born together, and have yet to go their separate ways. The fledgling stars have blundered into a cloud of gas in space, which looks like gossamer on webcam images. Even to the unaided eye or through binoculars, they are still a beautiful sight.

JANUARY'S PICTURE

The great **Orion Nebula** is lit by fierce radiation from a small cluster of newly-born stars called 'the Trapezium'. It's part of a huge gas complex, which has enough material to create half a million baby suns.

◄ *The Orion Nebula, photographed by Peter Shah using a 200 mm Newtonian reflector from Meifod, Powys. He gave 30-minute exposures through red, green and blue filters for the main nebula, plus separate 1-minute exposures in each colour for the central area, which would otherwise be overexposed.*

JANUARY'S TOPIC
Constellations

Canis Major, our constellation of the month, highlights humankind's obsession to 'join up the dots' in the sky, and weave stories around them – even if the shape of the star pattern bears little relation to its name.

But why? One answer is that the constellations on view change during the year as the Earth moves around the Sun, and the constellations acted as an '*aide memoire*' to where we are in our annual cycle – something of particular use to the ancient farming communities.

Another is that the stars were a great steer to navigation at sea. In fact, scholars believe that the Greek astronomers 'mapped' their legends on to the sky specifically so that sailors crossing the Mediterranean would associate certain constellations essential to navigation with their traditional stories.

But not all the world saw the sky through western eyes. The Chinese divided up the sky into a plethora of tiny constellations – with only three or four stars apiece. And the Australian Aborigines, in their pitch-black deserts, were so overwhelmed with stars that they made constellations out of the dark places where they couldn't see any stars!

The winter star-patterns are starting to drift towards the west, setting earlier – a sure sign that spring is on the way. The constantly changing pageant of constellations in the sky is proof that we live on a cosmic merry-go-round, orbiting the Sun. Imagine it: you're in the fairground, circling the Mighty Wurlitzer on your horse, and looking out around you. At times you spot the ghost train; sometimes you see the roller-coaster; and then you swing past the candy-floss stall. So it is with the sky – and the constellations – as we circle our local star. That's why we get to see different stars in different seasons.

FEBRUARY'S CONSTELLATION

Spectacular **Orion** is one of the rare star groupings that looks like its namesake – a giant of a man with a sword below his belt, wielding a club above his head. Orion is fabled in mythology as the ultimate hunter.

The constellation contains one-tenth of the brightest stars in the sky: its seven main stars all lie in the 'top 70' of brilliant stars. Despite its distinctive shape, most of these stars are not closely associated with each other – they simply line up, one behind the other.

Closest is the star that forms the hunter's right shoulder, **Bellatrix**, at 240 light years. Next is blood-red **Betelgeuse** at the top left of Orion, 600 light years away.

The constellation's brightest star, blue-white **Rigel**, is a vigorous young star more than twice as hot as our Sun, and 50,000 times as bright. Rigel lies 800 light years from us, roughly the same distance as the star that marks the other corner of Orion's tunic – **Saiph** – and the two outer stars of the belt, **Alnitak** (left) and **Mintaka** (right).

We must travel 1300 light years from home to reach the middle star of the belt, **Alnilam**. And at the same distance, we

▼ The sky at 10 pm in mid-February, with Moon positions at three-day intervals either side of Full Moon. The star positions are also correct for 11 pm at

e beginning of February, and
pm at the end of the month.
he planets move slightly relative
o the stars during the month.

find the stars of the 'sword' hanging below the belt – the lair of the great **Orion Nebula** (see Object and January's Picture).

PLANETS ON VIEW

Vying for attention with the giant of the constellations, Orion, we have the giant of the planets, **Jupiter**, putting on its last good evening show this year. Brilliant in the western sky at magnitude −2.0, in the dim constellation of Pisces, Jupiter sets at 9.25 pm at the start of February – but by the end of the month it's setting as early as 8.00 pm. Grab a pair of binoculars, and view the ever-changing pattern of the planet's four biggest moons as they orbit Jupiter, sometimes passing behind or in front of the celestial giant.

Uranus also lies in Pisces, glowing on the borderline of naked-eye visibility at magnitude +5.9. It's setting at 7.45 pm at the beginning of the month, but the planet sinks into invisibility in the twilight glow by the end of February.

Ringworld **Saturn**, at magnitude + 0.9, is rising at 10.40 pm when the month opens, and by 8.45 pm at the end of February. You'll find it in the constellation Virgo.

Last, but certainly not least, in the planetary parade is **Venus**, a celestial beacon at magnitude −4.1 in the south-eastern skies before dawn. The Morning Star is rising at 4.50 am at the start of February, but it's gradually dropping towards the Sun and rises at 5.10 am by month's end. Through a telescope, you'll see Venus gradually shrink in

WEST

PISCES
CETUS
TAURUS
ERIDANUS
PERSEUS
Pleiades
Aldebaran
LEPUS
12 Feb
Bellatrix
Mintaka
Alnilam
Rigel
ORION
Saiph
Mirzam
Betelgeuse
Alnitak
Orion Nebula
CANIS MAJOR
Capella
Sirius
Adhara
Zenith
AURIGA
GEMINI
15 Feb
Procyon
Castor
Pollux
CANIS MINOR
THE MILKY WAY
SOUTH
PUPPIS
URSA MAJOR
CANCER
The Sickle
Regulus
18 Feb
HYDRA
LEO
VIRGO
Saturn
Ecliptic
SE

EAST

February's Object
Orion Nebula

Saturn

Moon

MOON		
Date	**Time**	**Phase**
3	2.30 am	New Moon
11	7.18 am	First Quarter
18	8.36 am	Full Moon
24	11.26 pm	Last Quarter

◀ *Robin Scagell took this picture of Comet Hale-Bopp using a 35 mm camera with 135 mm telephoto lens mounted on a driven equatorial mount – there was no telescope involved. The exposure time was 5 minutes on ISO 1600 Ektachrome film on Good Friday, 29 March 1997, from a country location near Kingsbridge, Devon.*

size, while becoming more and more illuminated by the Sun.

Mercury, **Mars** and **Neptune** are all lost in the Sun's glare in February.

MOON

On 6 and 7 February, the crescent Moon lies near Jupiter, low in the evening sky. The First Quarter Moon passes the Seven Sisters (Pleiades) star cluster on 11 February. On the night of 20/21 February, the Moon lies below Saturn, and on 21/22 February it passes Spica. It's near another bright star, Antares, just before dawn on 25 February, and you'll find the very thin crescent Moon next to Venus just before the Sun rises on 28 February.

SPECIAL EVENTS

On **11 February**, the Moon lies just 1.5 degrees from the Pleiades star cluster, making a pretty grouping in the sky.

And on **14 February**, the Stardust space probe makes a flypast of Comet Tempel-1 (see this month's Topic).

FEBRUARY'S OBJECT

Below Orion's Belt lies a small fuzzy patch. Through binoculars, or a small telescope, the patch looks like a small cloud in space. It *is* a cloud – but at 30 light years across, it's hardly petite. Only the distance of the **Orion Nebula** – 1300 light years – diminishes it. Yet it is the nearest region to Earth where heavyweight stars are being born: this 'star factory' contains at least 150 fledgling stars (protostars), which have condensed out of dark clouds of gas and dust.

The Orion Nebula, illuminated by the searing radiation from young stars, is a great sight through a telescope or binoculars – and even with the unaided eye from a dark location.

FEBRUARY'S PICTURE

Comet Hale-Bopp, photographed on its spectacular appearance in spring 1997. The blue tail is caused by gas given off from the head, or nucleus, of the comet, while the yellowish tail is dust from the nucleus. A cousin to Tempel-1 – which the Stardust space probe will visit this month – this comet was one of the most sensational sky-sights in the past few years.

◉ Viewing tip

It may sound obvious, but if you want to stargaze at this most glorious time of year, dress up warmly! Lots of layers are better than a heavy coat, as they trap air next to your skin – and heavy-soled boots stop the frost creeping up your legs. It may sound anorakish, but a woolly hat really does stop one-third of your body's heat escaping through the top of your head. And, alas, no hipflask of whisky – alcohol constricts the veins, and makes you feel even colder.

FEBRUARY'S TOPIC
When Stardust meets Tempel-1...

Valentine's Day this year might be heavier on data, rather than on dates. On 14 February, the venerable Stardust space probe – which has been surveying the Solar System for over ten years – will get up close and personal to the comet Tempel-1. This icy piece of cosmic debris was the controversial victim of Deep Impact, a probe that explosively struck the comet with the intention of discovering its content and structure.

Stardust is no stranger to comets. In 2005, it visited the comet Wild 2, and collected dust samples which it returned to Earth. Now it will take images of the crater caused by Deep Impact and analyze the dust and gas that pour off this icy celestial wanderer, which orbits the Sun roughly every five years. The findings will prove to be interesting – for comets are the leftover building debris from the birth of our Solar System.

This month, the nights become shorter than the days as we hit the Vernal (Spring) Equinox – on 20 March, spring is 'official'. That's the date when the Sun climbs up over the Equator to shed its rays over the northern hemisphere. Because of the Earth's inclination of 23.5° to its orbital path around the Sun, the North Pole points away from our local star between September and March, causing the long nights of autumn and winter. Come the northern spring, Earth's axial tilt means that the Sun favours the north – and we can look forward to the long, warm days of summer.

Even better, the clocks 'spring forward' on 27 March, making daytime even longer.

MARCH'S CONSTELLATION

Canes Venatici (the Hunting Dogs) has to be one of the most obscure constellations ever invented. It lies directly under the tail of the Great Bear (**Ursa Major**).

Brought into the cosmos by Johannes Hevelius, the great 17th-century Polish astronomer (and member of a brewing circle), the pair of faint stars making up the constellation represent Asterion and Chara, two dogs chasing the Great Bear around the pole. Asterion is more familiarly known today as **Cor Caroli** – Charles' Heart – named by Sir Charles Scarborough after the doomed king Charles I of England.

The constellation houses a gem, although you'll need a fairly mighty telescope to see it as more than just a fuzzy blur. It's the glorious Whirlpool Galaxy (**M51**) – a face-on spiral with a little companion hand-in-hand with it (see Picture). In 1845, Lord Rosse, of Birr Castle in Ireland, sketched the pair as he saw them through his telescope. The 'Leviathan of Parsonstown' – then the largest astronomical instrument in the world – boasted a 1.8-metre (72-inch) mirror and it was the first to reveal spiral structure in galaxies.

▼ The sky at 10 pm in mid-March, with Moon positions at three-day intervals either side of Full Moon. The star positions are also correct for 11 pm at

...e beginning of March, and
...0 pm at the end of the month
...after BST begins). The planets
...move slightly relative to the stars
...during the month.

WEST

AURIGA

Aldebaran
TAURUS
ERIDANUS
Rigel
ORION
Betelgeuse
LEPUS
13 Mar
GEMINI
Castor
Pollux
Procyon
CANCER
CANIS MINOR
THE MILKY WAY
CANIS MAJOR
PUPPIS

URSA MAJOR
Zenith
The Sickle
Regulus
16 Mar
LEO
Moon
HYDRA
SOUTH

CANES VENATICI
Cor Caroli
M51
Denebola
19 Mar
CORVUS

BOÖTES
Arcturus
Saturn
VIRGO
Ecliptic
Spica

SERPENS

EAST

PLANETS ON VIEW

It's our last chance this year to see great **Jupiter** in the evening sky, brilliant at magnitude −1.9 in the constellation Cetus (just outside the zodiacal constellation Pisces). At the beginning of March, Jupiter is setting around 8 pm, but by the end of the month it has totally sunk into the twilight glow.

During the second half of March, Jupiter is joined in the evening sky by tiny **Mercury**, putting on its best evening appearance of 2011 – it's at greatest eastern elongation on 23 March. You'll spot Mercury very low in the west about an hour and a half after sunset. When the innermost planet first hoves into sight, around 12 March, it shines at magnitude −1.1, but it fades rapidly to magnitude +1.7 by the end of the month. At first, you'll see Mercury to the lower right of Jupiter; the planets pass on 15 March (see Special Events), and then Mercury lies directly above Jupiter.

Saturn (magnitude +0.7) lies in Virgo. It rises at 8.45 pm at the beginning of March, and at 7.45 pm at the end of the month.

Venus rises in the south-east an hour or so before the Sun. At magnitude −3.9, the brilliant Morning Star is hard to miss, even though it's now down in the twilight glow and won't be visible in a dark sky again until November.

Mars, **Uranus** and **Neptune** are too close to the Sun to be seen this month.

MOON

On the very first morning of the month, 1 March, the thin crescent Moon lies to

March's Object
The Moon

March's Picture
M51

Saturn

Moon

MOON		
Date	Time	Phase
4	8.46 pm	New Moon
12	11.45 pm	First Quarter
19	6.10 pm	Full Moon
26	12.07 pm	Last Quarter

the left of Venus. The evening of 6 March sees the waxing crescent Moon to the right of Jupiter. On 10 March, the Moon lies near the Pleiades, and on 17 March it passes Regulus. You'll find the Moon below Saturn and to the right of Spica on 20 March. On the mornings of 24 and 25 March the Moon is near Antares. And on the very last morning of the month, 31 March, the crescent Moon is back with Venus, low in the dawn sky.

SPECIAL EVENTS

On **15 March**, the Solar System's biggest world, Jupiter, lies next to the smallest planet, Mercury. Jupiter (to the left) appears about twice as bright as Mercury; in reality, it's almost 30 times wider, but it's currently over five times further away.

The Vernal (Spring) Equinox, on **20 March** at 11.21 pm, marks the beginning of spring, as the Sun moves up to shine over the northern hemisphere.

27 March, 1.00 am: British Summer Time starts – don't forget to put your clocks forward by an hour (the mnemonic is '*Spring* forward, *Fall* back').

MARCH'S OBJECT

The **Moon** is our nearest celestial companion, lying a mere 384,400 kilometres away. It took the Apollo astronauts only three days to reach it! And at 3476 kilometres across, it's so large when compared to Earth that – from space – the system would look like a double planet.

But the Moon couldn't be more different from our verdant Earth. Bereft of an atmosphere, it has been exposed to bombardment by meteorites and asteroids throughout its life. Even with the unaided eye, you can see the evidence. The 'face' of the 'Man in the Moon' consists of huge craters created by asteroid hits 3.8 billion years ago.

Through binoculars or a telescope, the surface of the Moon looks amazing – as if you're flying over it. But don't observe our satellite when it's Full: the light is

◉ Viewing tip

This is the time of year to tie down your compass points – the directions of north, south, east and west as seen from your observing site. North is easy – just latch on to Polaris, the Pole Star. And at noon, the Sun is always in the south. But the useful extra in March is that we hit the Spring Equinox, when the Sun rises due east, and sets due west. So remember those positions relative to a tree or house around your horizon.

flat and swamps its features. It's best to roam the Moon when it's a crescent and see the sideways-on shadows highlighting its dramatic relief.

MARCH'S PICTURE

M51, the Whirlpool Galaxy, is one of the most beautiful sky-sights – but it's the result of a cosmic traffic accident. Hundreds of millions of years ago, the small companion galaxy (seen above) collided with its larger neighbour, leading to a burst of starbirth in the Whirlpool. As a result, its glorious spiral arms are cradling young stars.

MARCH'S TOPIC
Messenger to the gods

Look out for elusive Mercury this month. Rumour has it that the architect of our Solar System – Nicolaus Copernicus – never observed the tiny world because of mists rising from the nearby River Vistula in Poland.

The pioneering space probe Mariner 10 sent back only fleeting images as it swung past the diminutive, cratered planet 35 years ago. But on 18 March, all is to change, for that's when NASA's Messenger probe goes into orbit around Mercury.

Messenger (the convoluted acronym stands for ME-rcury S-urface S-pace EN-vironment GE-ochemistry and R-anging mission) celebrates the belief that, in mythology, fleet-footed Mercury was messenger to the gods. Messenger has already made three flyhys of Mercury, to slow down the probe so that it can enter Mercury's orbit this month. By then, Messenger will have travelled almost 8 billion kilometres and made 15 circuits around the Sun.

Once in Mercury-orbit, Messenger's suite of instruments will scan the planet's surface and scrutinize its composition, offering a clue to the origins of this mysterious body – which may have been born further out from the Sun and later spiralled in. Messenger will also explore Mercury's internal workings – and it could confirm news of the latest discovery from Earth: that our innermost world has a core made of molten iron.

This enigmatic little planet will later be the target for the European probe BepiColombo, to be launched in 2013. BepiColombo comes in two parts: a Japanese spacecraft orbiting high above Mercury to probe its magnetism, and a European orbiter to examine its surface in extreme close-up.

◀ *Spiral galaxy M51 (the Whirlpool Galaxy) photographed by Eddie Guscott from Corringham, Essex, using a C9.25 Schmidt-Cassegrain telescope (235 mm aperture) on 25 March 2007. The total exposure time through separate colour filters was 3½ hours.*

The ancient constellations of **Leo** and **Virgo** dominate the springtime skies. Leo does indeed look like a recumbent lion, but it's hard to envisage Virgo as anything other than a vast 'Y' in the sky!

When you're looking at Virgo, spot the interloper. It's the planet Saturn, closest to the Earth this month. 'Close', however, is relative: the ringworld is over a billion kilometres distant.

▼ *The sky at 11 pm in mid-April, with Moon positions at three-day intervals either side of Full Moon. The star positions are also correct for midnight at the beginning of*

APRIL'S CONSTELLATION

The Y-shaped constellation of **Virgo** is the second-largest in the sky. It takes a bit of imagination to see the group of stars as a virtuous maiden holding an ear of corn (the bright star **Spica**), but this very old constellation has associations with the times of harvest. In the early months of autumn, the Sun passes through the stars of Virgo, hence the connections with the gathering-in of fruit and wheat.

Spica is a hot, blue-white star over 13,000 times brighter than the Sun, boasting a temperature of 22,500°C. It has a stellar companion, which lies just 18 million kilometres away from Spica – closer than Mercury orbits the Sun. Both stars inflict a mighty gravitational toll on each other, raising huge tides – thus creating two distorted, egg-shaped stars. Spica is the celestial equivalent of a rugby ball.

The glory of Virgo is the 'bowl' of the Y-shape. Scan it with a small telescope, and, you'll find it packed with faint, fuzzy blobs. These are just a few of the 2000 galaxies – star-cities like the Milky Way – that make up the gigantic **Virgo Cluster** (see Picture).

PLANETS ON VIEW

It's a pretty poor month for planet-lovers. Brilliant **Venus** (magnitude −3.8) is skulking in the dawn twilight: you may catch it just before sunrise low in the east.

April, and 10 pm at the end of the month. The planets move slightly relative to the stars during the month.

Otherwise, the only planet you'll see is **Saturn**. With the dark skies to itself, the ringworld is putting on its most magnificent display of the year, shining all night long at magnitude +0.6 in the constellation Virgo. On 3 April, Saturn is at opposition, closest to our planet – a 'mere' 1290 million kilometres away – and opposite the Sun as seen from Earth.

If you watch the planet carefully through a telescope from the beginning of April until the end of the month, you'll notice that the rings are significantly brighter around opposition. Saturn's gaudy appendages are made of icy particles that reflect light back towards the Sun (like motorway signs glaring in our car headlights), and this beam of reflected light is directed towards Earth when Saturn is opposite to the Sun in our skies.

Mercury, **Mars**, **Jupiter**, **Uranus** and **Neptune** are lost in the Sun's glare this month.

MOON

On 7 April, the crescent Moon lies near the Pleiades, low in the north-west. The Moon passes below Regulus on the night of 13/14 April, Saturn on 16 April and Spica on 17 April. You'll spot the Moon near Antares on the morning of 21 April. The thin crescent Moon lies above Venus just before dawn on 30 April.

SPECIAL EVENTS

22/23 April: It's the maximum of the **Lyrid** meteor shower, which – by perspective – appears to emanate from the constellation of Lyra. The

Star chart labels

WEST · THE MILKY WAY · GEMINI · Procyon · CANIS MINOR · Castor · Pollux · CANCER · 12 Apr · URSA MAJOR · Regulus · LEO · Denebola · HYDRA · Zenith · CANES VENATICI · M84/M86 · Virgo Cluster · 15 Apr · CORVUS · The Plough · Arcturus · VIRGO · Saturn · Ecliptic · Spica · SOUTH · CORONA BOREALIS · BOÖTES · SERPENS · 18 Apr · LIBRA · SW · HERCULES · OPHIUCHUS · SE · EAST

Saturn

Moon

April's Picture Virgo Cluster

MOON		
Date	Time	Phase
3	3.32 pm	New Moon
11	1.05 pm	First Quarter
18	3.44 am	Full Moon
25	3.47 am	Last Quarter

shower consists of particles from Comet Thatcher. It will be best to look for Lyrids after 2 am, when the Moon has set.

APRIL'S OBJECT

The **Sun** is livening up – and in more ways than one. As spring progresses, our local star climbs higher in the sky, and we feel the warmth of its rays. Some 150 million kilometres away, the Sun is our local star – and our local nuclear reactor.

This giant ball of hydrogen gas is a vast hydrogen bomb. At its core, where temperatures reach 15.7 million degrees, the Sun fuses atoms of hydrogen into helium. Every second, it devours 4 million tonnes of itself, bathing the Solar System with light and warmth.

But the Sun is also a dangerous place. It's now emerging from a quiet period to become active again; a cycle that repeats roughly 11 years. The driver is the Sun's magnetic field, wound up by the spinning of our star's surface gases. The magnetic activity suppresses the Sun's circulation, leading to a rash of dark sunspots. Then the pent-up energy is released in a frenzy of activity, when our star hurls charged particles through the Solar System. These dangerous particles can kill satellites, and even disrupt power lines on Earth. And if we are ever to make the three-year human journey to Mars, we will have to take the Sun's unpredictable, malevolent weather into account.

▲ *This photograph of the Virgo Cluster was taken by Eddie Guscott using a 130 mm refractor from Corringham, Essex, with a total exposure time of 5 hours 10 minutes through separate colour filters.*

Markarian's Chain, a line of galaxies in the **Virgo Cluster**, with the galaxies **M86** (centre) and **M84** (right). The Virgo Cluster of galaxies, 50 million light years away, dominates our place in the Universe. These huge, impressive galaxies – around 2000 of them – are the centre of our own local supercluster of galaxies. This massive agglomeration controls its galactic neighbourhood for millions of light years.

APRIL'S TOPIC
Saturn

The slowly moving ringworld **Saturn** is currently skulking in the peripheries of the sprawling constellation **Virgo** (the Virgin). It's famed for its huge engirdling appendages: the rings would stretch nearly all the way from the Earth to the Moon.

And the rings are just the beginnings of Saturn's larger family. It has at least 60 moons, including Titan – where the international Cassini-Huygens mission has discovered lakes of liquid methane and ethane. And the latest news is that Cassini has imaged plumes of salty water spewing from Saturn's icy moon Enceladus.

Saturn itself is second only to Jupiter in size. But it's so low in density that were you to plop it in an ocean, it would float. Like Jupiter, Saturn has a ferocious spin rate – 10 hours and 32 minutes – and its winds roar at speeds of up to 1800 km/h.

Saturn's atmosphere is much blander than that of its larger cousin. But it's wracked with lightning-bolts 1000 times more powerful than those on Earth.

◉ Viewing tip

Don't think that you need a telescope to bring the heavens closer. Binoculars are excellent – and you can fling them into the back of the car at the last minute. But when you buy binoculars, make sure that you get those with the biggest lenses, coupled with a modest magnification. Binoculars are described, for instance, as being '7×50' – meaning that the magnification is seven times, and that the diameter of the lenses is 50 mm across. These are ideal for astronomy – they have good light grasp, and the low magnification means that they don't exaggerate the wobbles of your arms too much. It's always best to rest your binoculars on a wall or a fence to steady the image. Some amateurs are the lucky owners of huge binoculars – say, 20×70 – with which you can see the rings of Saturn (being so large, these need a special mounting). But above all, never buy binoculars with small lenses that promise huge magnifications – they're a total waste of money.

Look up towards the south and you'll spot a distinctly orange-coloured star that lords it over a huge area of sky devoid of other bright stars. This is **Arcturus**, the brightest star in the constellation of **Boötes** (the Herdsman), who shepherds the two bears through the heavens. A sure sign that summer is on the way!

▼ The sky at 11 pm in mid-May, with Moon positions at three-day intervals either side of Full Moon. The star positions are also correct for midnight at the beginning of

MAY'S CONSTELLATION

Boötes (the Herdsman) is shaped rather like a kite. It was mentioned in Homer's Odyssey, and its name refers to the fact that Boötes seems to 'herd' the stars that lie in the northern part of the sky.

The name of the brightest star, **Arcturus**, means 'bear-driver'. It apparently 'drives' the Great Bear (**Ursa Major**) around the sky as the Earth rotates. Arcturus is the fourth brightest star in the whole sky, and it's the most brilliant star you can see on May evenings. A red giant star (see Topic) in its old age, Arcturus lies 37 light years from us, and shines 110 times more brilliantly than the Sun.

The star at the ten o'clock position from Arcturus is called **Izar**, whose name means 'the belt'. Through a good telescope, it appears as a gorgeous double star – one star is yellow and the other blue.

PLANETS ON VIEW

The spring 2011 planetary drought continues into May. The noble exception is **Saturn**, still strutting its stuff in the constellation Virgo. Visible all night long in the south, the ringworld shines at magnitude +0.8, and is gradually moving towards Porrima, the star marking the centre of the Y-shaped constellation.

Telescope-users can now spot distant **Neptune** (magnitude +7.9) in Aquarius, rising in the south-east around 2.30 am.

May, and 10 pm at the end of the month. The planets move slightly relative to the stars during the month.

Venus is skulking low on the south-eastern horizon an hour before sunrise, but even at a brilliant magnitude −3.8 you'll need to know exactly where to look to catch it in the dawn twilight.

It will be equally difficult to spot **Jupiter**, when it begins to rise from the sunrise glow in the last few days of May. Making its debut an hour and a half before the Sun, the giant planet shines at magnitude −2.0 in Pisces, lying to the right of Venus.

Mercury (at western elongation on 7 May), **Mars** and **Uranus** are too close to the Sun to be seen this month.

MOON

The Moon lies near Regulus on 11 May. On 13 May, it passes below Saturn, and on 14 May the Moon is near Spica. On 17 and 18 May, the star keeping the Moon company low in the south is Antares. The crescent Moon lies immediately above Jupiter just before dawn on 29 May, and next to Venus on the morning of 31 May.

SPECIAL EVENTS

The maximum of the Eta Aquarid meteor shower falls on **4/5 May**, when tiny pieces of Halley's Comet burn up in Earth's atmosphere. Because the Moon is well out of the way, 2011 is a good year for observing these shooting stars.

MAY'S OBJECT

At the darkest part of a May night, you may spot a faint fuzzy patch in **Hercules**, up high in the

WEST

EAST

May's Object
M13

May's Picture
The Plough

Saturn

Moon

MOON		
Date	**Time**	**Phase**
3	7.51 am	New Moon
10	9.33 pm	First Quarter
17	12.09 pm	Full Moon
24	7.52 pm	Last Quarter

south. Through binoculars, it appears as a gently glowing ball of light. With a telescope, you can glimpse its true nature: a cluster of almost a million stars, swarming together in space.

This wonderful object is known as **M13**, because it was the 13th entry in the catalogue of fuzzy objects recorded by the 18th-century French astronomer Charles Messier. We now classify M13 as a 'globular cluster'. These great round balls of stars are among the oldest objects in our Milky Way Galaxy, dating back to its birth some 13 billion years ago.

In 1974, radio astronomers sent a message towards M13, hoping to inform the inhabitants of any planet there of our existence. There's only one problem: M13 lies so far away that we wouldn't receive a reply until AD 52,200!

MAY'S PICTURE

The Plough is our best-loved constellation – always visible, always circling the Pole Star. Children call it 'the Saucepan' – to Americans it's 'the Big Dipper'. And earlier generations dubbed it 'Charles' Wain'. Officially, it's part of **Ursa Major** (the Great Bear). Look out for the double star, **Mizar** and **Alcor**, in the bend of the bear's tail (see June's Object).

◉ Viewing tip

When you first go out to observe, you may be disappointed at how few stars you can see in the sky. But wait for around 20 minutes, and you'll be amazed at how your night vision improves. One reason for this 'dark adaption' is that the pupil of your eye gets larger to make the best of the darkness. More importantly, in dark conditions the retina of your eye builds up much bigger reserves of rhodopsin, the chemical that responds to light.

You can use the two 'end' stars of the Plough (**Merak** and **Dubhe**) to locate **Polaris**. Draw a line between these two 'pointers', and extend it to hit the Pole Star (see chart on page 22).

MAY'S TOPIC
Red giants

Arcturus, the fourth-brightest star in the sky (magnitude −0.04) always gladdens our hearts as the harbinger of summer's arrival. But, alas, it is a star on the way out. Its orange colour indicates that it's a red giant: a star near the end of its life.

Arcturus is about 30 times wider than the Sun - even though it comes in at the same weight – because it has swollen up in its old age. This is a fate that not only befalls humans, it also happens to stars (although the causes are rather different!). Stars generate energy by nuclear fusion: they 'burn' hydrogen into the next-up element, helium, in their hot cores.

But there's only a finite supply of hydrogen in a star's core. When it runs out – after around 10 billion years – the core, made of helium, collapses to a smaller size and gets even hotter. The star is now like an onion, with distinct layers. In the centre, the compressed helium switches on its own reactions. And there's a layer around the core, where hydrogen is still turning to helium. These reactions create more heat, causing the outer layers of the star to balloon in size and cool down – hence the baleful red or orange colour of the star's surface.

Eventually Arcturus, like all red giants, will lose the grip on its atmosphere and jettison it into space, as a glowing planetary nebula (see June's Topic). All that's left will be the dying core: a steadily cooling object, about the size of the Earth, called a white dwarf. This fate awaits our Sun – but not for another 7 billion years.

◀ *The Plough, photographed on Ektachrome 1600 film by Robin Scagell. He used a 1-minute exposure through a 58 mm lens with a diffusion filter in front which expanded the larger star images, bringing out their colour.*

This month, our local star reaches its highest position over the northern hemisphere, when we get the longest days and shortest nights. The Summer Solstice this year takes place on 21 June, and the height of summer will no doubt be celebrated at festivals around the northern hemisphere on this day – notably at Stonehenge, in Wiltshire. This seasonal ritual traces its roots back through millennia, and has led to the construction of massive stone monuments aligned on the rising Sun at midsummer. Our ancestors clearly had formidable astronomical knowledge.

JUNE'S CONSTELLATION

Lyra is one of the most perfect little constellations in the sky. Shaped like a Greek lyre, it's dominated by brilliant white **Vega**, the fifth-brightest star in the sky. Just 25 light years away – a near-neighbour in the Cosmos – Vega is surrounded by a disc of dust that has probably given birth to baby planets.

Next to Vega is **epsilon Lyrae**, a quadruple star known as the 'double-double'. Keen-sighted people can separate the pair, but you'll need a small telescope to find that each star is itself double.

The gem of Lyra lies between the two end stars of the constellation, **beta** and **gamma Lyrae**. This one needs a serious telescope, as it's nearly 9th magnitude. The **Ring Nebula** is a cosmic smoke-ring: the end of the life for a star like the Sun. We explore these beautiful, but grim, cosmic corpses on page 29 (see Topic).

PLANETS ON VIEW

Saturn lords it over the evening sky, at magnitude +0.9 in Virgo. All month long, the ringed planet lies very near Porrima, the star at the centre of the constellation's

▼ The sky at 11 pm in mid-June, with Moon positions at three-day intervals either side of Full Moon. The star positions are also correc for midnight at the beginning of

WEST

Ecliptic

Regulus

6 June

The Sickle

CANCER

Pollux

LEO

Castor

GEMINI

URSA MAJOR

CANES VENATICI

AURIGA

Dubhe

The Plough

Alkaid

Mizar/ Alcor

Zenith

HERCULES

Capella

URSA MINOR

DRACO

NORTH

Polaris

M 57

Vega

CASSIOPEIA

CEPHEUS

epsilon LYRA

gamma

PERSEUS

Algol

THE MILKY WAY

Deneb

CYGNUS

ANDROMEDA

DELPHINUS

NE

Square of Pegasus

PEGASUS

EAST

une, and 10 pm at the end of the month. The planets move lightly relative to the stars during the month.

Y-shape. Soon after Saturn sets in the west around 2 am, a brighter challenger rises in the east: **Jupiter** (magnitude −2.0), blazing among the faint stars of Pisces and Aries.

The two more distant giant planets are also to be found in June's night sky. **Neptune** (magnitude +7.9) rises about 0.30 am: this telescopic world is inhabiting Aquarius. At magnitude +5.9, **Uranus** is technically visible to the naked eye – but certainly not in June's bright summer nights! Uranus lies in Pisces and rises about 1.30 am.

Venus is still horizon-hugging in the dawn twilight, a lone light in the brightening sky, at magnitude −3.8. And, in the last week of June, **Mars** emerges at long last from the glare of the Sun's light. You'll find the Red Planet in Taurus, rising in the north-east just before 3 am – although at only magnitude +1.3 it's not yet making much of a statement.

Mercury is the only planet too close to the Sun to be seen in June.

MOON

On 7 June, the waxing Moon lies near Regulus, while 10 June sees the Moon passing below Saturn. On 11 June, the Moon is near Spica. The almost Full Moon lies right above Antares on 14 June. In the early hours of 26 June, the crescent Moon is next to Jupiter. Just before dawn on 28 and 29 June, the narrow crescent is near Mars (the Pleiades are just above, but will be difficult to spot in the twilight).

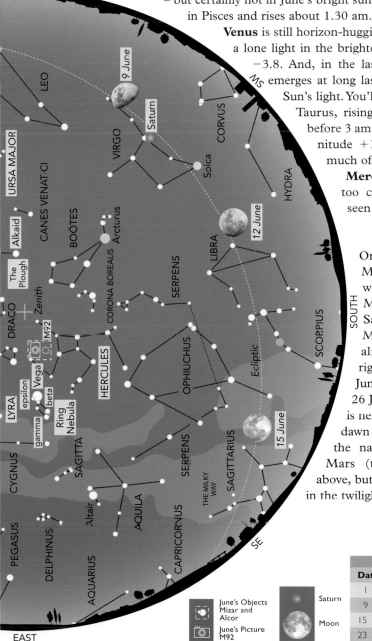

MOON		
Date	Time	Phase
1	10.02 pm	New Moon
9	3.11 am	First Quarter
15	9.13 pm	Full Moon
23	11.48 pm	Last Quarter

June's Objects
Mizar and Alcor

June's Picture
M92

Saturn

Moon

27

SPECIAL EVENTS

There's a partial eclipse of the Sun on **1 June**, but you'll need to be somewhere around the Arctic Ocean to see anything of it.

21 June, 6.16 pm: Summer Solstice. The Sun reaches its most northerly point in the sky, so 21 June is Midsummer's Day, with the longest period of daylight. Correspondingly, we have the shortest nights.

We're due for an eclipse of the Moon on **15 June**, total as seen from the UK – but unfortunately not in a dark sky. The Moon is totally eclipsed when it rises in the south-east at 9.20 pm, but, with the bright dusk sky, you won't see it at all. The Moon starts to move out of the Earth's shadow at 10.00 pm, and returns to Full Moon glory by 11.00 pm.

▲ Globular cluster M92 in Hercules, photographed by Eddie Guscott using a 130 mm refracting telescope from Corringham, Essex, on 28 April 2006. The total CCD exposure time through separate red, green and blue colour filters was 2 hours 20 minutes.

JUNE'S OBJECT

Home in on the 'kink' in the tail of **Ursa Major** (the Great Bear), and you'll spot the most famous pair of stars in the sky – **Mizar** (magnitude +2.4) and **Alcor** (magnitude +4.0). Generations of astronomers have referred to them as 'the horse and rider', and students have been raised on the fact that the pair make up a classic double star system, orbiting in each other's embrace. But *are* Mizar and Alcor an item? It seems not. Although they both lie about 80 light years away, they are separated by 3 light years – nearly the distance from the Sun to our closest star, Proxima Centauri. Undoubtedly, Mizar is a

◉ *Viewing tip*

This is the month for maximum Sun-viewing – but be careful. **NEVER** use a telescope or binoculars to look at the Sun directly: it could blind you permanently. Fogged film is no safer, because it allows the Sun's infra-red (heat) rays to get through. Eclipse goggles are safe (unless they're scratched). The best way to observe the Sun is to project its image through binoculars or a telescope on to a white piece of card.

complex star system, having a companion visible through a telescope which is itself double; in total, there are four stars involved. But it appears that Alcor is an innocent bystander. Although it shares its path through space with Mizar, the two are just members of most of the stars that make up the 'stellar association' of Ursa Major. Unlike most constellations, the stars of the Great Bear are genuinely linked by birth (with the exceptions of **Dubhe** and **Alkaid**, at opposite ends of the central '**Plough**'). So let's hear 'independence for Alcor'!

JUNE'S PICTURE

M92 is a globular cluster of some 330,000 stars located at the top of the constellation of **Hercules**. It is overshadowed by its brighter neighbour M13 (see May's Object), but it is nevertheless a lovely cluster. In crystal-clear skies, M92 is just visible to the unaided eye. It was discovered in 1777 by Johann Bode. Lying 26,000 light years away, M92 is one of an estimated 158 globular clusters – all made of ancient stars – that live in the outer 'halo' of our Galaxy.

JUNE'S TOPIC
Planetary nebulae

William Herschel – who discovered the planet Uranus – first named these fuzzy objects 'planetary nebulae', because, to him, they looked like the planet that he had found. Now we know that they are the end-point for the majority of stars.

When a star like the Sun runs out of its nuclear fuel at the end of its life, its core – in a desperate attempt to keep the star alive – contracts, heating up its outer layers. The star becomes unstable and puffs off its atmosphere into space in the shape of a ring.

The ring disperses in a few thousand years. It leaves the derelict core inside to leak away its energy – ending up as a cold, lonely black dwarf.

The most famous planetary nebula in the sky is the **Ring Nebula** in **Lyra** – a classic cosmic smoke-ring.

High summer is here, and with it comes the brilliant trio of the **Summer Triangle** – the stars **Vega**, **Deneb** and **Altair**. Each is the brightest star in its own constellation: Vega in **Lyra**, Deneb in **Cygnus**, and Altair in **Aquila**. And this is the time to catch the far-southern constellations of **Sagittarius** and **Scorpius** – embedded in the glorious heart of the Milky Way.

▼ *The sky at 11 pm in mid-July, with Moon positions at three-day intervals either side of Full Moon. The star positions are also correct for midnight at the beginning of*

JULY'S CONSTELLATION

Low down in the south, you'll find a constellation that's shaped rather like a teapot – **Sagittarius**. The handle lies to the left and the spout to the right!

To the ancient Greeks, the star-pattern of Sagittarius represented an archer, with the torso of a man and the body of a horse. The 'handle' of the teapot represents his upper body, the curve of three stars to the right are his bent bow, while the end of the spout is the point of the arrow, aimed at **Scorpius**, the fearsome celestial scorpion.

Sagittarius is rich with nebulae and star clusters. If you have a clear night (and preferably from a southern latitude), sweep Sagittarius with binoculars for some fantastic sights. Above the spout lies the wonderful **Lagoon Nebula** – visible to the naked eye on clear nights. This is a region where stars are being born. Between the teapot and the neighbouring constellation **Aquila**, you'll find a bright patch of stars in the Milky Way (catalogued as **M24**). Raise your binoculars higher to spot another star-forming region, the **Omega Nebula**.

Finally, on a very dark night you may spot a fuzzy patch, above and to the left of the teapot's lid. This is the globular cluster **M22**, a swarm of almost a million stars that lies 10,000 light years away.

WEST

NW

LEO

The Sickle

VIRGO

BOÖTES

CANES VENATICI

The Plough

HERCULES

URSA MAJOR

DRACO

AURIGA

URSA MINOR

Polaris

Zenith

CYGNUS

NORTH

Capella

Deneb

CASSIOPEIA

CEPHEUS

THE MILKY WAY

North America Nebula

PERSEUS

Algol

PEGASUS

TRIANGULUM

ANDROMEDA

Square of Pegasus

NE

PISCES

EAST

ly, and 10 pm at the end of
e month. The planets move
ightly relative to the stars
uring the month.

PLANETS ON VIEW

The planetary scene is fairly humming again, after the doldrums of the past few months. During the first week of July, you may just catch **Mercury** (magnitude −0.1) very low in the north-west around 10.30 pm; it reaches greatest elongation on 20 July, past the time of best visibility.

During the evening, **Saturn** is the major planet on view. Shining at magnitude +1.0 in Virgo, close to the star Porrima, it's now setting around midnight.

Neptune (magnitude +7.8) lies in Aquarius and rises about 10.30 pm. It's followed by slightly brighter **Uranus**, just on naked-eye visibility at magnitude +5.8, rising at around 11.30 pm in Pisces.

If you wait up until 1 am, you can't miss glorious **Jupiter** rising in the east in the neighbouring constellation of Aries. The giant planet is the brightest object in July's night sky (after the Moon), at magnitude −2.2.

Mars rises just after 2 am, and you may catch it low in the north-east before the sky brightens. The Red Planet lies in Taurus, and shines at magnitude +1.4.

Rising just an hour before the Sun, at around 4 am, **Venus** (magnitude −3.8) is very difficult to spot in the dawn twilight.

MOON

On 3 July, the thin crescent Moon lies to the left of Mercury – in the twilight sky, the view is best with binoculars. On 7 July, the Moon is below Saturn, and on 8 July

Star chart labels

WEST · 6 July · Saturn · VIRGO · Spica · 9 July · LIBRA · Arcturus · BOÖTES · CORONA BOREALIS · SERPENS · OPHIUCHUS · 12 July · Antares · SCORPIUS · SERPENS · DRACO · Zenith · HERCULES · SAGITTA · LYRA · Vega · SERPENS · M24 · Omega Nebula · Lagoon Nebula · SOUTH · CYGNUS · Deneb · SUMMER TRIANGLE · THE MILKY WAY · M22 · 15 July · SAGITTARIUS · North America Nebula · Altair · AQUILA · CAPRICORNUS · PEGASUS · DELPHINUS · Neptune · Ecliptic · SE · PISCES · 18 July · AQUARIUS · EAST

Moon / Object key

□ July's Object — Summer Triangle
📷 July's Picture — North America Nebula

● Saturn
● Neptune
● Moon

MOON		
Date	Time	Phase
1	9.54 am	New Moon
8	7.29 am	First Quarter
15	7.39 am	Full Moon
23	6.02 am	Last Quarter
30	7.40 pm	New Moon

the First Quarter Moon passes Spica. The Moon is near Antares on 11 and 12 July. In the morning of 24 July, the Moon is immediately above Jupiter. The crescent Moon lies near Mars on the mornings of 27 and 28 July.

SPECIAL EVENTS

There's a tiny partial eclipse of the Sun on **1 July**, but you need to be near Antarctica to see it at all!

On **4 July**, the Earth is at aphelion, its furthest point from the Sun.

JULY'S OBJECT

The **Summer Triangle** is very much part of this season's skies (and it hangs around for most of the autumn, too!). It's made up of Vega, Deneb and Altair – the brightest stars in the constellations of Lyra, Cygnus and Aquila, respectively. The trio of stars make a striking pattern almost overhead on July nights.

The stars may seem to be almost the same brightness, but they're very different beasts. **Altair** – its name means 'flying eagle' – is one of the Sun's nearest neighbours, at a distance of nearly 17 light years. It's about ten times brighter than the Sun and spins at a breakneck rate of once every 6.5 hours – as compared to around 30 days for our local star.

Vega, just over 25 light years away, is a brilliant white star nearly twice as hot as the Sun. In 1850, it was the first star to be photographed. Now, more sensitive instruments have revealed that Vega is surrounded by a dusty disc, which may be a planetary system in the process of formation.

While **Deneb** – meaning 'tail' (of the swan) – may appear to be the faintest of the trio, the reality is different. It lies a staggering 3200 light years away (a newly-measured distance from the Hipparcos satellite). To appear so bright in our skies, it must be truly luminous. We now know that Deneb is over 200,000 times brighter than our Sun – one of the most brilliant stars known.

JULY'S PICTURE

The aptly-named **North America Nebula**, with its companion, the Pelican Nebula, lies next to brilliant Deneb in Cygnus. They are actually part of a single enormous cloud of gas and dust, poised to create generations of new stars. The gap between them is caused by intervening dark interstellar dust. The nebula glows red, the result of heating from a nearby star. Researchers suspect that it could well be Deneb. If this is the case, the nebulae are about 1800 light years away – and the North America Nebula measures a staggering 100 light years across.

▶ *Philip Perkins photographed the North America Nebula from Wiltshire through a 400 mm telephoto lens mounted on top of a driven telescope. He combined four separate 40-minute exposures on Ektapress film.*

◉ *Viewing tip*

This is the month when you really need a good, unobstructed view to the southern horizon to make out the summer constellations of Scorpius and Sagittarius. They never rise high in temperate latitudes, so make the best of a southerly view – especially over the sea – if you're away on holiday. A good southern horizon is also best for views of the planets, because they rise highest when they're in the south.

JULY'S TOPIC
DAWN at Vesta

This may seem like *déjà vu*: yes, we did feature the DAWN mission to the asteroids in *Stargazing 2006*. But the mission was postponed, cancelled, and then reinstated. This month, however, the daring NASA mission will reach its first target, the brightest asteroid, Vesta; the second is Ceres, the biggest asteroid (930 kilometres across). It will orbit each asteroid for a year or more, to check them out at close quarters. The mission's aim is to investigate the formation and evolution of our Solar System. Asteroids and comets are the building-blocks of planets, and give us clues to our origins – perhaps even the origin of life. The two asteroids couldn't be more different, which is why NASA has chosen them as targets. Ceres evolved with water present – there may even be frost or water vapour on its surface today. Vesta, on the other hand, originated in hot and violent circumstances. By studying the two asteroids, DAWN will be looking back to the earliest days of our Solar System.

Well, it should have been the Glorious Twelfth for astronomers this month: 12/13 August is the maximum of the **Perseid** meteor shower. But this year it's drowned out by moonlight. However – never fret! You can use the warm nights to take in a fine display of summer stars, and – this year – a great gathering of planets in the west.

▼ The sky at 11 pm in mid-August, with Moon positions at three-day intervals either side of Full Moon. The star positions are also correct for midnight

AUGUST'S CONSTELLATION

The cosmic dragon writhes between the two bears in the northern sky. **Draco** is probably associated with Hercules' 12 labours, because its head rests on the (upside-down) superhero's feet. In this case, Hercules had to get past a crowd of nymphs and slay a 100-headed dragon (called Ladon) before completing his task – which, in this case, was stealing the immortal golden apples from the gardens of the Hesperides.

Confusingly, the brightest star in Draco is gamma Draconis – which ought to be third in the pecking order. Also known as **Eltanin**, this orange star shines at magnitude +2.2 and lies 148 light years away. But all this is to change: in 1.5 million years, it will swing past the Earth at a distance of 28 light years, outshining even Sirius.

Alpha Draconis (also known as **Thuban**), which by rights should be the brightest in the constellation, stumbles in at a mere magnitude +3.7. Thuban lies just below Ursa Minor, in the tail of the dragon, 300 light years away.

But what Thuban lacks in brightness, it makes up for in fame. Thuban was our Pole Star in the years around 2800 BC. It actually lay closer to the celestial pole than Polaris does now – just 2.5 arcminutes, as opposed to 42 for Polaris.

The swinging of Earth's axis – like the toppling of a spinning top – takes place over a period of 26,000 years. 'Precession'

WEST

CANES VENATICI
Arcturus
BOÖTES
CORONA BOREALIS
The Plough
URSA MAJOR
HERCULES
Thuban
DRACO
URSA MINOR
Polaris
Eltanin
Zenith
CYGNUS
Deneb
NORTH
CASSIOPEIA
CEPHEUS
PEGASUS
Capella
Radiant of Perseids
THE MILKY WAY
AURIGA
PERSEUS
ANDROMEDA
Algol
TRIANGULUM
19 Aug
Jupiter
ARIES
Pleiades
CETUS
PISCES
Ecliptic

NE

EAST

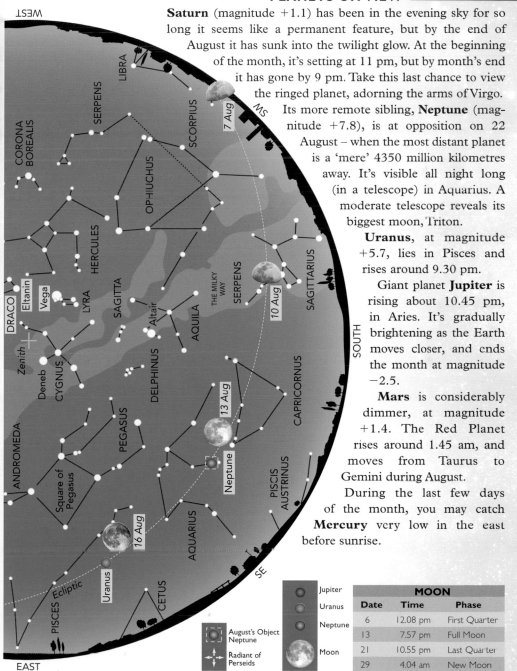

at the beginning of August, and 10 pm at the end of the month. The planets move slightly relative to the stars during the month.

means that the Earth's north pole points to a number of stars over the millennia, which we spin 'underneath' – so they appear stationary in the sky. Look forward to AD 14,000, when brilliant **Vega** will take over the pole!

PLANETS ON VIEW

Saturn (magnitude +1.1) has been in the evening sky for so long it seems like a permanent feature, but by the end of August it has sunk into the twilight glow. At the beginning of the month, it's setting at 11 pm, but by month's end it has gone by 9 pm. Take this last chance to view the ringed planet, adorning the arms of Virgo.

Its more remote sibling, **Neptune** (magnitude +7.8), is at opposition on 22 August – when the most distant planet is a 'mere' 4350 million kilometres away. It's visible all night long (in a telescope) in Aquarius. A moderate telescope reveals its biggest moon, Triton.

Uranus, at magnitude +5.7, lies in Pisces and rises around 9.30 pm.

Giant planet **Jupiter** is rising about 10.45 pm, in Aries. It's gradually brightening as the Earth moves closer, and ends the month at magnitude −2.5.

Mars is considerably dimmer, at magnitude +1.4. The Red Planet rises around 1.45 am, and moves from Taurus to Gemini during August.

During the last few days of the month, you may catch **Mercury** very low in the east before sunrise.

		Jupiter
		Uranus
		Neptune

August's Object
Neptune

Radiant of Perseids

Moon

MOON		
Date	**Time**	**Phase**
6	12.08 pm	First Quarter
13	7.57 pm	Full Moon
21	10.55 pm	Last Quarter
29	4.04 am	New Moon

Venus is lost in the Sun's glare in August.

▲ *A Perseid meteor, photographed on 12 August 2007 from St Lawrence Bay, Essex, by Bob Purbrick. He mounted his digital camera on a tripod and pointed it at the stars in the hope that a meteor would appear during the 30-second exposure time. This was the only meteor that appeared in 59 such exposures on the same night. It was his first attempt at meteor photography!*

MOON

The crescent Moon passes Saturn on 3 August, and Spica on 4 August. Antares is the star near the Moon on 7 and 8 August. On 19 and 20 August, you'll find the waning Moon near Jupiter. The Moon passes below the Pleiades on 21 August. In the mornings of 25 and 26 August, the crescent Moon is near Mars. The thin crescent lies to the upper right of Mercury just before dawn on 27 August.

SPECIAL EVENTS

The maximum of the annual **Perseid** meteor shower falls on **12/13 August**, but this is a very poor year for observing them as brilliant moonlight will drown out all but the brightest meteors.

AUGUST'S OBJECT

With Pluto having been demoted to the status of being a mere 'ice dwarf', **Neptune** is officially the most remote planet in our Solar System. It lies about 4500 million kilometres away from the Sun – 30 times the Earth's distance – in the twilight zone of our family of worlds. The gas giant takes nearly 165 years to circle the Sun. It has just completed one 'Neptune year' since its discovery in 1846.

This month, Neptune is just visible through a small telescope low in Aquarius and at its closest this year on 22 August. But you need a space probe to get up close and personal to the planet. In 1989, Voyager 2 managed this. It discovered a turquoise world 17 times more massive than Earth, cloaked in clouds of methane and ammonia.

Neptune has a family of 13 moons – one of which, Triton, boasts erupting ice volcanoes. And, like the other outer worlds, it's encircled by rings of debris – although, in Neptune's case, they're very faint.

For a planet so far from the Sun, Neptune is amazingly frisky. Its core blazes at 7000°C – hotter than the surface of the Sun. This internal heat drives dramatic storms, and winds of 2000 km/h – the fastest in the Solar System.

AUGUST'S PICTURE

Early August is usually a great time to see shooting stars. The **Perseids** – debris from Comet Swift-Tuttle – stream into the atmosphere and burn up at the rate of roughly one a minute. This year, bright moonlight will swamp the cosmic fireworks display, but we won't be able to overlook a meteor as bright as this one!

AUGUST'S TOPIC
Juno and NuSTAR missions

This month, two major launches are taking place to monitor very different environments in the Universe. On 5 August, NASA's Juno probe blasts off from Cape Canaveral in Florida. This solar-powered craft is bound for Jupiter, the biggest world in our Solar System. It will orbit the giant planet's poles 33 times, to investigate the gas-world's atmosphere, structure and awesome magnetic field.

Just ten days later – on 15 August – NASA will launch NuSTAR, a ground-breaking mission to search for the most extreme results of violence in the Universe. It will sniff out black holes, map the gruesome results of supernova explosions, and study the most devastating effects of exploding galaxies. Watch this space – from behind the sofa!

The month's wet weather is most likely with us again – and we have the star-patterns to match. **Aquarius** (the Water Carrier) is part of a group of aqueous star-patterns that include **Cetus** (the Sea Monster), **Capricornus** (the Sea Goat), **Pisces** (the Fishes), **Piscis Austrinus** (the Southern Fish) and **Delphinus** (the Dolphin). There's speculation that the ancient Babylonians associated this region with water because the Sun passed through this zone of the heavens during their rainy season, from February to March.

SEPTEMBER'S CONSTELLATION

Cetus – the sky's pet sea monster – straggles below the barren square of **Pegasus**. Like its fellow constellation, it has very few stars of note – bar one. **Mira** ('The Wonderful') was discovered in 1596 by the astronomer David Fabricius.

Over an 11-month period, it varies in brightness alarmingly. Normally, it changes between magnitude +3 and magnitude +10, but in 1779 William Herschel observed it as being as bright as Arcturus, the fourth brightest star in the sky.

Mira is a distended red giant, unable to control its girth. At its faintest, it is only about as bright as our Sun, but it can swell to become 1500 times brighter. This 'star out of hell' is 700 times bigger than our local star, and would engulf all the planets as far out as Jupiter if it were to be placed in our Solar System. Its fate will be to puff off its unstable atmosphere, and turn into a planetary nebula (see June's Topic).

Mira should reach maximum in the last week of September – but bear in mind that variable stars are as predictable as cats!

▼ The sky at 11 pm in mid-September, with Moon positions at three-day intervals either side of Full Moon. The star positions are also correct for midnight at

PLANETS ON VIEW

Jupiter is now rising in the east, in Aries, just as the sky grows

*he beginning of September, and
*0 pm at the end of the month.
The planets move slightly relative
to the stars during the month.*

dark, around 8.45 pm. At magnitude −2.6, the brilliant planet dominates the view all night long. Use binoculars or a small telescope to track the ever-changing pattern of its four brightest moons.

At magnitude +7.8, you'll certainly need optical aid to view **Neptune**. The most distant planet lies in the constellation Aquarius and sets at about 4.30 am.

Its slightly brighter twin, **Uranus** (magnitude +5.7), is at opposition on 25 September and is above the horizon all night. Technically visible to the naked eye – under perfectly dark and clear skies – it's more easily found in binoculars or a small telescope.

Mars rises about 1.20 am, at magnitude +1.3. During September, Mars moves from Gemini to Cancer, and at the end of the month it approaches the Praesepe star cluster.

In the dawn skies, you may catch elusive **Mercury** during the first half of September, very low in the east around 5 am. The tiny planet is at greatest western elongation on 3 September, and it brightens from magnitude +0.3 to −1.1 during the first two weeks of the month. On the morning of 9 September, Mercury passes close to Regulus.

Venus and **Saturn** lie too close to the Sun to be seen this month.

MOON

On 4 September, the First Quarter Moon lies above Antares. The Moon is near Jupiter on 16 September. The morning of 23 September sees the waning crescent Moon next to Mars;

	MOON	
Date	**Time**	**Phase**
4	6.39 pm	First Quarter
12	10.26 am	Full Moon
20	2.39 pm	Last Quarter
27	12.09 pm	New Moon

September's Object
Uranus

September's Picture
Veil Nebula

on the morning of 25 September, it's just below Regulus.

SPECIAL EVENTS

On **8 September**, the GRAIL spacecraft heads off from Cape Canaveral towards the Moon. The Gravity Recovery and Interior Laboratory mission will investigate our satellite's structure from crust to core, and probe the evolution of our Moon.

It's the Autumn Equinox at 10.04 am on **23 September**. The Sun is over the Equator as it heads southwards in the sky, and day and night are equal.

SEPTEMBER'S OBJECT

If you're very sharp-sighted and have extremely dark skies, you stand a chance of spotting **Uranus** – the most distant planet visible to the unaided eye. At magnitude +5.7 in Pisces, it's

▼ *The Veil Nebula in Cygnus, photographed through Peter Shah's 200 mm Newtonian reflector. He gave a total of 50 minutes' exposure time through red, green and blue filters.*

◉ *Viewing tip*

It's best to view your favourite objects when they're well clear of the horizon. When you look low down, you're seeing through a large thickness of the atmosphere – which is always shifting and turbulent. It's like trying to observe the outside world from the bottom of a swimming pool! This turbulence makes the stars appear to twinkle. Low-down planets also twinkle – although to a lesser extent because they subtend tiny discs, and aren't so affected.

closest to Earth this year on 25 September. Discovered in 1781 by amateur astronomer William Herschel, Uranus was the first planet to be found since antiquity. Then, it doubled the size of our Solar System. Uranus is a gas giant like Jupiter, Saturn and Neptune. Four times the diameter of the Earth, it has an odd claim to fame: it orbits the Sun on its side (probably as a result of a collision in its infancy, which knocked it off its perch). Like the other gas giants, it has an encircling system of rings. But these are nothing like the spectacular edifices that girdle Saturn: the 11 rings are thin and faint. It also has a large family of moons – at the last count, 27. Many of us were disappointed when the Voyager probe flew past Uranus in 1986 to reveal a bland, featureless world. But things are hotting up as the planet's seasons change. Streaks and clouds are appearing in its atmosphere.

SEPTEMBER'S PICTURE

The haunting wraiths of the **Veil Nebula** in **Cygnus** are the remains of a star that exploded over 5000 years ago. The twisted filaments are part of the Cygnus Loop – a reservoir of material swept up by the dead star that is expanding into space. One day, a phoenix will rise from the ashes: the gas will create a new generation of stars and planets – and, possibly, life.

SEPTEMBER'S TOPIC
The Harvest Moon

It's the time of year when farmers work late into the night, bringing home their ripe crops before autumn sets in. And traditionally they are aided by the light of the 'Harvest Moon' – a huge glowing Full Moon that seems to hang constant in the evening sky, rising at almost the same time night after night. At first sight, that doesn't seem possible. After all, the Moon is moving around the Earth, once in just under a month, so it ought to rise roughly one hour later every night. But things in the sky are hardly ever that simple....

The Moon follows a tilted path around the sky (close to the line of the Ecliptic, which is marked on the chart). And this path changes its angle with the horizon at different times of the year. On September evenings, the Moon's path runs roughly parallel to the horizon, so night after night the Moon moves to the left in the sky, but it hardly moves downwards. As a consequence, the Moon rises around the same time for several consecutive nights. This year, Full Moon on 12 September rises at 6.55 pm (ideal for harvesting) – it rises just 17 minutes earlier the evening before, and 17 minutes later the night after.

Jupiter is king of the night sky. To its right, you'll find the barren **Square of Pegasus** riding high in the south, joined by the faint and undistinguished constellations of autumn.

▼ *The sky at 11 pm in mid-October, with Moon positions at three-day intervals either side of Full Moon. The star positions are also correct for midnight at*

OCTOBER'S CONSTELLATION

It takes considerable imagination to see the line of stars making up **Andromeda** as a young princess chained to a rock, about to be gobbled up by a vast sea monster (**Cetus**) – but that's ancient legends for you. Despite its rather mundane appearance, the constellation contains some surprising delights. One is **Almach**, the star at the left-hand end of the line. It's a beautiful double star. The main star is a yellow supergiant shining 650 times brighter than the Sun, and its companion – which is 5th magnitude – is bluish. The two stars are a lovely sight in small telescopes. Almach is actually a quadruple star: its companion is in fact triple.

But the glory of Andromeda is its great galaxy, beautifully placed on October nights. Lying above the line of stars, the **Andromeda Galaxy** (see Picture) is the most distant object easily visible to the unaided eye. It lies a mind-boggling 2.5 million light years away, yet it's so vast that it appears nearly four times as big as the Full Moon in the sky (although the sky will seldom be clear enough to allow you to see the faint outer regions).

The Andromeda Galaxy is the biggest member of the Local Group – it's estimated to contain over 400 billion stars. It is a wonderful sight in binoculars or a small telescope, and the latter will reveal its two bright companion galaxies – M32 and NGC205.

PLANETS ON VIEW

Giant planet **Jupiter** is the undoubted 'star' of the month (see Object)! It's at opposition on 29 October, when it's opposite to the Sun in our skies and at its closest to Earth this

*he beginning of October, and
pm at the end of the month
after the end of BST). The planets
move slightly relative to the stars
during the month.*

year. This enormous world is visible in the southern sky all night long, at magnitude −2.8, totally outshining the surrounding stars of Aries. A small telescope shows Jupiter's flattened globe, with multicoloured bands of cloud, while even binoculars will reveal the perpetual dance of its four biggest moons.

Meanwhile, competition is beginning to emerge as **Venus** (magnitude −3.8) begins to make its presence felt in the evening sky. By the end of October – while still in the twilight glow – the Evening Star is setting in the south-west almost an hour after the Sun.

Mars, at magnitude +1.2, rises in the north-east around 1 am. It starts the month in Praesepe (see Special Events). During October the Red Planet moves from Cancer to Leo.

Telescopic **Neptune** (magnitude +7.9) lies on the borders of Capricornus and Aquarius. The most distant planet sets around 2.30 am. **Uranus**, at magnitude +5.7, is in Pisces and sets about 5.30 am.

Mercury and **Saturn** are lost in the Sun's glare in October.

MOON

On 1 October, the crescent Moon lies just above Antares. The Moon passes Jupiter on 13 October. In the morning of 22 October, you'll find the crescent Moon below Mars, with Regulus to its left.

SPECIAL EVENTS

In the early hours of **1** and **2 October**, you'll find Mars right in front of Praesepe, the star cluster in

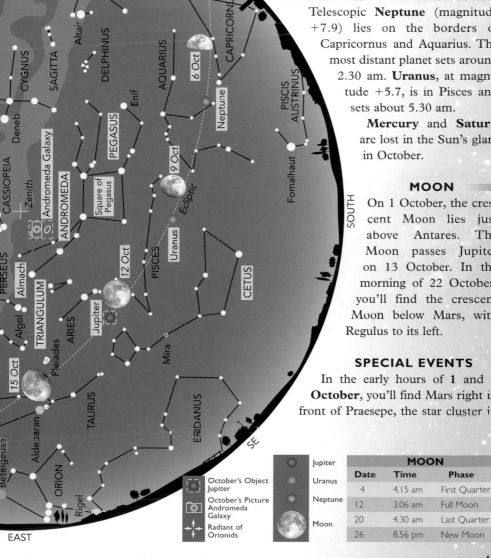

	MOON		
	MOON		
Date	**Time**	**Phase**	
4	4.15 am	First Quarter	
12	3.06 am	Full Moon	
20	4.30 am	Last Quarter	
26	8.56 pm	New Moon	

Jupiter
October's Object Jupiter
Uranus
October's Picture Andromeda Galaxy
Neptune
Radiant of Orionids
Moon

Cancer known as the Beehive – a lovely sight in binoculars or a small telescope.

Debris from Halley's Comet smashes into Earth's atmosphere on **21/22 October**, causing the annual **Orionid** meteor shower.

At 2 am on **30 October**, we see the end of British Summer Time for this year. Clocks go backwards by an hour.

OCTOBER'S OBJECT

Jupiter is particularly bright this month. On 29 October, it's at opposition – meaning that it's opposite the Sun in the sky, and at its closest to the Earth. 'Close' is a relative term, however – the planet is still over 590 million kilometres away. But Jupiter is so vast (at 143,000 kilometres in diameter, it could contain 1300 Earths) and as it's made almost entirely of gas, it's very efficient at reflecting sunlight.

Although Jupiter is so huge, it spins faster than any other planet in the Solar System. It rotates every 9 hours 55 minutes, and as a result its equator bulges outwards – through a small telescope, it looks a bit like a tangerine crossed with an old-

fashioned humbug. The humbug stripes are cloud belts of ammonia and methane stretched out by the planet's dizzy spin.

Jupiter has a fearsome magnetic field that no astronaut would survive, huge lightning storms, and an internal heat source which means it radiates more energy than it receives from the Sun. Jupiter's core simmers at a temperature of 20,000°C.

Jupiter commands its own 'mini-solar system' – a family of over 60 moons. The four biggest are visible in good binoculars, and even – to the really sharp-sighted – to the unaided eye. These are worlds in their own right (Ganymede is even larger than the planet Mercury). But two vie for 'star' status: Io and Europa. The surface of Io has incredible geysers erupting plumes of sulphur dioxide 300 kilometres into space. Brilliant white Europa probably contains oceans of liquid water beneath a solid ice crust, where alien fish may swim....

OCTOBER'S PICTURE

The **Andromeda Galaxy** is one of the biggest spiral galaxies known, containing more stars than the Milky Way. Alas, it is presented to us at such a shallow angle that even the best photographs can't show the true glory of spiral arms. In this image, its companion galaxies – M32 (top) and NGC 205 (bottom) – flank their giant host.

◄ *The Andromeda Galaxy, M31, photographed by Ian King using a 75 mm Pentax refractor and Starlight Xpress SXV-M25 CCD camera. He used separate exposures through individual colour filters.*

OCTOBER'S TOPIC
Mars missions

America, Russia and China are all hoping to blast off to Mars this month. NASA is planning to launch its planetary rover Curiosity to the Red Planet. Ten times heavier than the current rovers on the Martian surface, this awesome beast will scoop up the planet's soil, and determine if it was – or, today, is – capable of harbouring microbial lifeforms.

Meanwhile, the Russians are busying themselves with the wonderfully named Phobos Grunt, aimed at Phobos – the Red Planet's major moon. Piggy-backing on board will be the Chinese Yinghuo-1 orbiting space probe. The two will part on arrival at Mars.

Phobos Grunt means 'Phobos soil'. And that's exactly what the spacecraft is designed to investigate. It will dig up a soil sample from Phobos – an irregularly-shaped moon less than 30 kilometres across – and send it back to Earth in 2012. Its composition will tell researchers how the Martian system came into being. It may also answer the question as to whether Phobos is a genuine moon of the Red Planet – or a planetoid that Mars has snatched from the nearby asteroid belt.

'A swarm of fireflies tangled in a silver braid' was the evocative description of the **Pleiades** star cluster by Alfred, Lord Tennyson, in his 1842 poem 'Locksley Hall'. All over the world, people have been intrigued by this lovely sight. From Greece to Australia, ancient myths independently describe the stars as a group of young girls being chased by an aggressive male – often **Aldebaran** or **Orion** – giving rise to the cluster's popular name, the Seven Sisters. Polynesian navigators used the Pleiades to mark the start of their year. And farmers in the Andes rely on the visibility of the Pleiades as a guide to planting their potatoes: the brightness or faintness of the Seven Sisters depends on El Niño, which affects the forthcoming weather. But once you see the Pleiades, you know that winter is here.

NOVEMBER'S CONSTELLATION

It has to be said that **Pegasus** is one of the most boring constellations in the sky. A large, barren square of four medium-bright stars – how did our ancestors manage to see the shape of an upside-down winged horse up there?

In legend, Pegasus sprang from the blood of Medusa the Gorgon when **Perseus** (nearby in the sky) severed her head. But all pre-classical civilizations have their fabled winged horse, and we see them depicted on Etruscan and Euphratean vases.

The star at the top right of the square – **Scheat** – is a red giant over a hundred times wider than the Sun. Close to the end of its life, it pulsates irregularly, changing in brightness by about a magnitude. **Enif** ('the nose') – outside the square to the lower right – is a yellow supergiant. A small telescope, or even good binoculars, will reveal a faint blue companion star.

Just next to Enif – and Pegasus' best-kept secret – is the beautiful globular cluster **M15**. You'll need a telescope for this

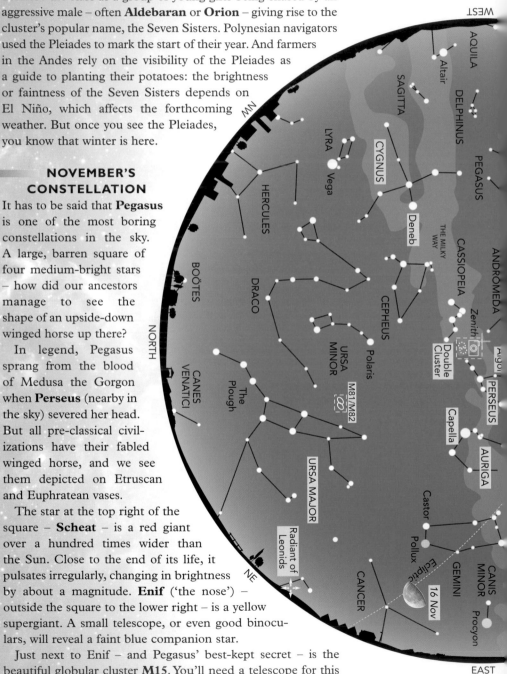

▼ The sky at 10 pm in mid-November, with Moon positions at three-day intervals either side of Full Moon. The star positions are also correct for 11 pm at

he beginning of November, and
pm at the end of the month.
he planets move slightly relative
o the stars during the month.

one. M15 is around 50,000 light years away, and contains about 200,000 stars.

PLANETS ON VIEW

There's action low in the west after sunset, as **Venus** moves rapidly upwards. By the end of November, you'll find the Evening Star blazing at magnitude −3.8 in the south-west, and setting one-and-a-half hours after the Sun.

From 10 to 15 November, you may just catch **Mercury** (magnitude −0.1) lying below Venus. (It reaches greatest elongation on 14 November.) Because of the twilight glow, the innermost planet is best seen in binoculars.

Once Venus has set, **Jupiter** reigns supreme, at magnitude −2.7, on the borders of Aries and Pisces. The giant planet sets around 5.30 am. At magnitude +7.9, a small telescope shows **Neptune** in Aquarius, setting around 11.30 pm. Binoculars will suffice to reveal **Uranus** (magnitude +5.7): lying in Pisces, the dim green planet sets at about 2.30 am.

Mars brightens from magnitude +1.1 to +0.7 during November, as the Earth catches up with the slower planet. Rising about 11.30 pm, the Red Planet lies in Leo. It passes the lion's brightest star, Regulus, on 9 November: compare the ochre colour of Mars with the blue-white star.

And just before dawn, you can spot **Saturn** returning from its sojourn behind the Sun. The ringed planet is rising around 4.30 am, in

WEST

AQUILA
DELPHINUS
CYGNUS
M15
Enif
Neptune
CAPRICORNUS
PEGASUS
4 Nov
AQUARIUS
Fomalhaut
Scheat
Square of Pegasus
Uranus
Ecliptic
7 Nov
Andromeda Galaxy
CASSIOPEIA
ANDROMEDA
PISCES
CETUS
Zenith
Algol
TRIANGULUM
ARIES
Jupiter
Mira
Double Cluster
PERSEUS
Pleiades
10 Nov
TAURUS
ERIDANUS
Capella
AURIGA
Aldebaran
ORION
LEPUS
SOUTH
13 Nov
Betelgeuse
Rigel
CANIS MINOR
THE MILKY WAY
Procyon
SE

EAST

		MOON	
	Date	**Time**	**Phase**
	2	4.38 pm	First Quarter
	10	8.16 pm	Full Moon
	18	3.09 pm	Last Quarter
	25	6.10 am	New Moon

November's Objects M81/M82
November's Picture Double Cluster
Radiant of Leonids

Jupiter
Uranus
Neptune
Moon

Virgo. At magnitude +1.0 it's a close twin to the constellation's brightest star, Spica, which lies just below Saturn.

MOON

You'll find the Moon, almost Full, near Jupiter on 9 November. On the morning of 19 November, the waning Moon lies below Mars and Regulus. And the mornings of 22 and 23 November see the crescent Moon near Saturn and Spica. Back in the evening sky, the waxing crescent Moon forms a lovely duo with Venus on 26 and 27 November.

SPECIAL EVENTS

The night of **17/18 November** sees the maximum of the **Leonid** meteor shower. A few years ago, this annual shower yielded literally storms of shooting stars, but the rate has gone down as the parent Comet Tempel-Tuttle, which sheds its dust to produce the meteors, has moved away from the vicinity of Earth.

There's a partial eclipse of the Sun on **25 November**, but it's only visible from Antarctica and parts of the Southern Ocean.

NOVEMBER'S OBJECT

A pair of galaxies this month, in **Ursa Major**, called **M81** and **M82**. You can just see each of these galaxies with binoculars, on a really dark night, though a moderately powerful telescope is needed to reveal them in detail. 'M', incidentally, stands for Charles Messier, an 18th-century Parisian astronomer who catalogued 103 'fuzzy objects' that misleadingly resembled comets, which he was desperate to discover. He did find a handful of comets – but today, he's better remembered for his Messier Catalogue.

M81 is a beautiful, smooth spiral galaxy, like our Milky Way, with spiral arms wrapped around a softly-glowing core. It's similar in size and mass to our own Galaxy, and lies 11 million light years away.

Lying close by is M82 – another spiral galaxy, but one which couldn't be more different. It looks a total mess, with a huge eruption taking place at its core. This is a result of an interaction with M81 some 300 million years

▼ *The Double Cluster in Perseus, NGC 869 (right) and NGC 884, photographed by Peter Shah through a 200 mm Orion Optics AG8 reflector with Starlight Xpress CCD camera, with four separate 10-minute exposures – one each through red, green and blue filters, and an unfiltered exposure.*

Viewing tip

Now that the nights are drawing in earlier, and becoming darker, it's a good time to pick out faint, fuzzy objects like the Andromeda Galaxy and the Orion Nebula. But don't even think about it near the time of Full Moon – its light will drown them out. The best time to observe 'deep-sky objects' is when the Moon is near to New, or well after Full Moon. Check the Moon phases timetable in the book.

ago, when the two galaxies pulled streams of interstellar gas out of one another. Gas clouds are still raining on to M82's core, creating an explosion of star formation. For this reason, M82 is called a 'starburst' galaxy.

NOVEMBER'S PICTURE

This beautiful pair of star clusters in **Perseus** – the **Double Cluster** – is a glorious sight in binoculars. Medium-sized telescopes reveal that each cluster contains about 300 stars, but this is only the tip of the iceberg: there are probably thousands of stars in residence. The stars are very young (in astronomical terms!) – between 3 million and 5 million years old, as compared to our Sun's 5000 million years.

NOVEMBER'S TOPIC
Star names

Why do the brightest stars have such strange names? The reason is that they date from antiquity, and have been passed on down generations ever since. The original western star names – like the original constellations – were probably Babylonian or Chaldean, but few of these survive. The Greeks took up the baton after that, and the name of the star **Antares** is a direct result. It means 'rival of Ares' because its red colour rivals that of the planet Mars (*Ares* in Greek).

The Romans were not particularly interested in astronomy, but nevertheless left their mark on the sky. **Capella**, the brightest star of **Auriga**, has Roman roots: the name is a diminutive of *capra* (goat), and it literally means 'the little she-goat' (a bit of an understatement for a star over 100 times brighter than the Sun).

But the Arabs were largely responsible for the star names we have inherited today. Working in the so-called 'Dark Ages' between the 6th and 10th centuries AD, they took over the naming of the sky – hence the number of stars beginning with the letters 'al' (Arabic for 'the'). **Algol**, in the constellation **Perseus**, means 'the demon' – possibly because the Arabs noticed that its brightness seems to 'wink' every few days. **Deneb**, in **Cygnus**, also has Arabic roots – it means 'the tail' (of the flying bird).

But the most remembered star name in the sky is Orion's **Betelgeuse**. For some time, it was gloriously interpreted as 'the armpit of the sacred one'. But the 'B' in Betelgeuse turned out to be a mistransliteration – and so we're none the wiser as to how our distant ancestors really identified this fiery red star.

This year, we have a Christmas Star: the planet Venus, shining brilliantly in the sunset at twilight. But it's the nadir of the year – December sees the shortest day, and the longest night, when Earth's North Pole is turned away from our local star. On 22 December, we hit the Winter Solstice. It has long been commemorated in tablets of stone, aligned to welcome the returning Sun.

▼ The sky at 10 pm in mid-December, with Moon positions at three-day intervals either side of Full Moon. The star positions are also correct for 11 pm at

DECEMBER'S CONSTELLATION

Taurus is very much a second cousin to brilliant **Orion**, but a fascinating constellation none-theless. It's dominated by **Aldebaran**, the baleful blood-red eye of the celestial bull. Around 68 light years away, and shining with a magnitude of +0.85, Aldebaran is a red giant star, but not one as extreme as neighbour-ing **Betelgeuse**. It is around three times heavier than the Sun. The 'head' of the bull is formed by the **Hyades** star cluster. The other famous star cluster in Taurus is the far more glamorous **Pleiades**, whose stars – although further away than the Hyades – are younger and brighter.

Taurus has two 'horns' – the star **El Nath** (Arabic for 'the butting one') to the north, and **zeta Tauri** (whose Babylonian name Shurnarkabti-sha-shutu, meaning 'star in the bull towards the south', is thankfully not generally used!). Above this star is a stellar wreck – literally. In 1054, Chinese astronomers witnessed a brilliant 'new star' appear in this spot, which was visible in daytime for weeks. What the Chinese actually saw was an exploding star – a supernova – in its death throes. And today, we see its still-expanding remains as the **Crab Nebula**. It's visible through a medium-sized telescope.

PLANETS ON VIEW

Venus is the Christmas Star of 2011. The Evening Star hangs like a lantern in the south-west after sunset, at magnitude −3.9. By New Year's Eve, it's setting three hours after the Sun and is visible in a dark sky for the first time since January.

Venus's brilliance has knocked **Jupiter** from its dominance. You can find the giant planet at 'only' magnitude −2.5 (still brighter than any star) on the borders of Aries and Pisces, and setting around 3.15 am.

Faint **Neptune** (magnitude +7.9), in Aquarius, is setting around 9.30 pm. It's followed by slightly brighter twin **Uranus** (magnitude +5.8), lying in Pisces and dropping below the horizon about 0.30 am.

After months hiding itself away until the early hours, **Mars** is now part of the evening scene, rising in the east about 10.45 pm (just too late for our chart). During the month, the Red Planet tracks along under the belly of Leo, and brightens from magnitude +0.7 to +0.2.

Saturn rises around 2.40 am, shining at magnitude +1.0 in Virgo, to the left of Spica.

Finally, in the second part of the month you may catch **Mercury** (magnitude 0.0) in the dawn twilight; it's at greatest western elongation on 23 December. Look low in the south-east around 6.30 am, between 12 and 31 December.

MOON

On 6 December, the Moon lies immediately above Jupiter. On the night of

the beginning of December, and 9 pm at the end of the month. The planets move slightly relative to the stars during the month.

Star chart labels

WEST · EAST · SOUTH · SE · MS

AQUARIUS · PEGASUS · Square of Pegasus · ANDROMEDA · PISCES · Ecliptic · Uranus · CETUS · Mira · TRIANGULUM · ARIES · Jupiter · Algol · PERSEUS · Pleiades · Hyades · TAURUS · ERIDANUS · Zenith · Capella · AURIGA · El Nath · Crab Nebula · zeta · Aldebaran · Rigel · LEPUS · ORION · Betelgeuse · Horsehead Nebula · COLUMBA · GEMINI · Castor · Pollux · Radiant of Geminids · CANIS MINOR · Procyon · THE MILKY WAY · Sirius · CANIS MAJOR · Adhara · CANCER · HYDRA

1 Dec · 4 Dec · 7 Dec · 10 Dec · 13 Dec

	December's Object Algol
	December's Picture Horsehead Nebula
	Radiant of Geminids

Jupiter · Uranus · Moon

MOON		
Date	**Time**	**Phase**
2	9.52 am	First Quarter
10	2.37 pm	Full Moon
18	0.47 am	Last Quarter
24	6.06 pm	New Moon

10/11 December, the Full Moon moves directly in front of the Crab Nebula (see Special Events). The Moon passes below Regulus on 15 December, and it's near Mars on 16 and 17 December. On the morning of 20 December, the crescent Moon lies below Saturn and Spica. The dawn skies of 22 and 23 December see the narrow crescent near to Mercury. On the evenings of 26 and 27 December, there's a lovely pairing of the waxing crescent and Venus.

▶ Peter Shah took this view of the Horsehead Nebula in Orion using his 200 mm Newtonian reflector and CCD. He gave a total exposure time of 2 hours 50 minutes through separate red, green, blue and hydrogen alpha filters from his home in Meifod, Wales. North is at the left.

SPECIAL EVENTS

The maximum of the **Geminid** meteor shower falls on **13/14 December**. These shooting stars are debris shed from an asteroid called Phaethon and therefore quite substantial – and hence bright. But this year the show will be spoilt by moonlight.

There's a total eclipse of the Moon on **10 December**, but it will be a damp squib as seen from the UK. The total phase is over by the time the Moon rises in the north-east at 4.00 pm, and the partial eclipse finishes by 4.15 pm.

On the night of **10/11 December** (between 2.30 and 3.30 am), the Moon occults the Crab Nebula. Even with a telescope, this unusual event will be a challenge, as the Moon will be just past Full, and at its brightest.

The Winter Solstice occurs at 5.30 am on **22 December**. As a result of the tilt of Earth's axis, the Sun reaches its lowest point in the heavens as seen from the northern hemisphere: we get the shortest days, and the longest nights.

DECEMBER'S OBJECT

The star **Algol**, in the constellation **Perseus**, represents the head of the dreadful Gorgon Medusa. In Arabic, its name means 'the demon'. Watch Algol carefully and you'll see why. Every 2 days 21 hours, Algol dims in brightness for several hours – to become as faint as the star lying to its lower right (Gorgonea Tertia).

In 1783, a young British amateur astronomer, John Goodricke of York, discovered Algol's regular changes, and proposed that Algol is orbited by a large dark planet that periodically blocks off some of its light. We now know that Algol does indeed have a dim companion blocking its brilliant light, but it's a fainter star rather than a planet.

DECEMBER'S PICTURE

One of the most iconic of astronomical tourist spots, the **Horsehead Nebula** is part of the huge star-forming region in the constellation of **Orion**. This celestial chess-piece measures 4 light years from 'nose' to 'mane'. It's a dark cloud of cosmic soot, which will collapse under gravity to form a new generation of stars.

◉ Viewing tip

This is the month when you may be thinking of buying a telescope as a Christmas present for a budding stargazer. Beware! Unscrupulous mail-order catalogues selling 'gadgets' often advertise small telescopes that boast huge magnifications. This is known as 'empty magnification' – blowing up an image that the lens or mirror simply doesn't have the ability to get to grips with, so all you see is a bigger blur. A rule of thumb is to use a maximum magnification no greater than twice the diameter of your lens or mirror in millimetres. So if you have a 100 mm reflecting telescope, go no higher than 200X.

DECEMBER'S TOPIC
Search for Extra-Terrestrial Intelligence

Whenever we give a presentation, one of the questions we're most asked is: 'Is there anybody out there?' It's now half a century since the veteran American astronomer Frank Drake turned his radio telescope to the heavens in the hope of hearing an alien broadcast. Despite false alarms (probably caused by secret military equipment), there has been a deafening silence.

Drake and his colleagues founded an independent institution in California – the SETI Institute (Search for Extra-Terrestrial Intelligence). It's a serious scientific endeavour that looks into the biology, psychology and motivation for our search for alien life.

Recently, their fortunes have been boosted by a donation from Paul Allen (who co-founded Microsoft). Thanks to Allen, the team is now building an array of 400 radio telescopes in California to tune in to that first whisper from ET.

But is life out there far more advanced than us? Is radio communication something that came and went? The SETI researchers are contemplating communicating with laser beams – but even that may prove too primitive.

There's always something to see in our Solar System, from planets to meteors or the Moon. These objects are very close to us – in astronomical terms – so their positions, shapes and sizes appear to change constantly. It is important to know when, where and how to look if you are to enjoy exploring Earth's neighbourhood. Here we give the best dates in 2011 for observing the planets and meteors (weather permitting!), and explain some of the concepts that will help you to get the most out of your observing.

THE INFERIOR PLANETS

A planet with an orbit that lies closer to the Sun than the orbit of Earth is known as *inferior*. Mercury and Venus are the inferior planets. They show a full range of phases (like the Moon) from the thinnest crescents to full, depending on their position in relation to the Earth and the Sun. The diagram below shows the various positions of the inferior planets. They are invisible when at *conjunction*, when they are either behind the Sun, or between the Earth and the Sun, and lost in the latter's glare.

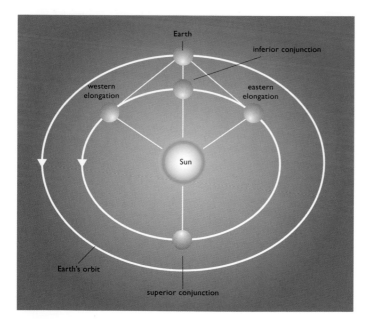

Mercury

In the first few weeks of January, Mercury is visible low in the south-east before dawn; it reaches its greatest western elongation on 9 January. It re-emerges in the second half of March for its best evening appearance of the year, but disappears again in April. Its best morning appearances are in early September and mid–late December.

Magnitudes

Astronomers measure the brightness of stars, planets and other celestial objects using a scale of *magnitudes*. Somewhat confusingly, fainter objects have higher magnitudes, while brighter objects have lower magnitudes; the most brilliant stars have negative magnitudes! Naked-eye stars range from magnitude −1.5 for the brightest star, Sirius, to +6.5 for the faintest stars you can see on a really dark night.

As a guide, here are the magnitudes of selected objects:

Sun	−26.7
Full Moon	−12.5
Venus (at its brightest)	−4.7
Sirius	−1.5
Betelgeuse	+0.4
Polaris (Pole Star)	+2.0
Faintest star visible to the naked eye	+6.5
Faintest star visible to the Hubble Space Telescope	+31

◀ At eastern or western elongation, an inferior planet is at its maximum angular distance from the Sun. Conjunction occurs at two stages in the planet's orbit. Under certain circumstances, an inferior planet can transit across the Sun's disc at inferior conjunction.

⬤ Maximum elongations of Mercury in 2011

Date	Separation
9 January	23.3° west
23 March	18.6° east
7 May	26.6° west
20 July	26.8° east
3 September	18.1° west
14 November	22.7° east
23 December	21.8° west

Maximum elongation of Venus in 2011	
Date	Separation
8 January	47.0° west

Venus

Venus reaches its greatest western elongation of 2011 on 8 January. From the beginning of the year onwards it is brilliant in the east as a Morning Star, hugging the horizon, but disappears from view by August. It re-emerges as an Evening Star towards the end of October.

THE SUPERIOR PLANETS

The superior planets are those with orbits that lie beyond that of the Earth. They are Mars, Jupiter, Saturn, Uranus and Neptune. The best time to observe a superior planet is when the Earth lies between it and the Sun. At this point in a planet's orbit, it is said to be at *opposition*.

▶ *Superior planets are invisible at conjunction. At quadrature the planet is at right angles to the Sun as viewed from Earth. Opposition is the best time to observe a superior planet.*

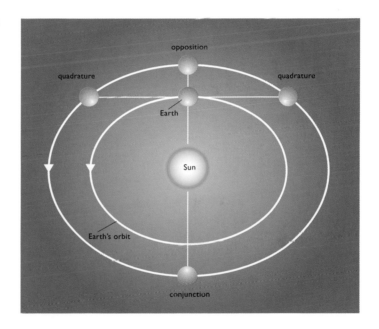

Progress of Mars through the constellations	
Late June – end July	Taurus
August – mid-Sept	Gemini
Mid-Sept – mid-Oct	Cancer
Mid-Oct – end Dec	Leo

Mars

Mars is lost from view at the start of the year, and only emerges from the Sun's glare in the last week of June. The Red Planet is at its best during the second half of the year, reaching its brightest magnitude of 2011 in December.

Jupiter

Jupiter is visible in the west from January onwards, in Pisces, but by the end of March it has already disappeared into the twilight glow. It starts to reappear in the east at the very end of May. Moving into Aries, Jupiter brightens each month until its best showing in October, when it is visible in the southern sky all night long. It reaches opposition on 29 October.

Saturn

Saturn is in Virgo all year. It is at opposition on 3 April, and is at its best for observing from January to July. It is lost from view by the end of August, but re-emerges in the morning during November.

Uranus

Uranus is best viewed in early January, when it lies in Pisces, but by the end of February it has disappeared into the twilight glow. It reappears in June, and visibility improves into the autumn. Uranus reaches opposition on 25 September.

Neptune

Neptune is best viewed from May onwards, when it lies in Aquarius. It is at opposition on 22 August.

SOLAR AND LUNAR ECLIPSES

Solar Eclipses

There are four partial solar eclipses in 2011. The first, on 4 January, can be seen from south-east England. However, the second partial eclipse, on 1 June, will only be viewable from around the Arctic Ocean. The third partial eclipse, on 1 July, can be seen only from near Antarctica; similarly, the fourth partial eclipse of the year, on 25 November, will only be viewable from Antarctica and parts of the Southern Ocean.

Lunar Eclipses

The lunar eclipse on 15 June will be total as seen from the UK, but unfortunately not in a dark sky, so it will not be visible after all. The total lunar eclipse on 10 December will also be a disappointment for UK observers, as the total phase is over by the time the Moon rises in the north-east.

> **Astronomical distances**
> For objects in the Solar System, such as the planets, we can give their distances from the Earth in kilometres. But the distances are just too huge once we reach out to the stars. Even the nearest star (Proxima Centauri) lies 25 million million kilometres away.
>
> So astronomers use a larger unit – the *light year*. This is the distance that light travels in one year, and it equals 9.46 million million kilometres.
>
> Here are the distances to some familiar astronomical objects, in light years:
>
> | | |
> |---|---:|
> | Proxima Centauri | 4.2 |
> | Betelgeuse | 600 |
> | Centre of the Milky Way | 26,000 |
> | Andromeda Galaxy | 2.5 million |
> | Most distant galaxies seen by the Hubble Space Telescope | 13 billion |

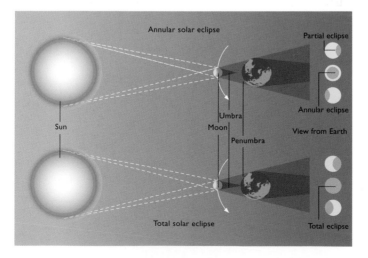

◀ Where the dark central part (the umbra) of the Moon's shadow reaches the Earth, we see a total eclipse. People located within the penumbra see a partial eclipse. If the umbral shadow does not reach the Earth, we see an annular eclipse. This type of eclipse occurs when the Moon is at a distant point in its orbit and is not quite large enough to cover the whole of the Sun's disc.

▶ *Meteors from a common source, occurring during a shower, enter the atmosphere along parallel trajectories. As a result of perspective, however, they appear to diverge from a single point in the sky – the radiant.*

METEOR SHOWERS

Shooting stars – or *meteors* – are tiny particles of interplanetary dust, known as *meteoroids*, burning up in the Earth's atmosphere. At certain times of year, the Earth passes through a stream of these meteoroids (usually debris left behind by a comet) and we see a *meteor shower*. The point in the sky from which the meteors appear to emanate is known as the *radiant*. Most showers are known by the constellation in which the radiant is situated.

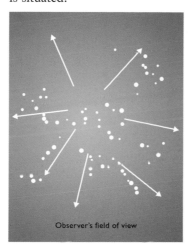

Observer's field of view

When watching meteors for a co-ordinated meteor programme, observers generally note the time, seeing conditions, cloud cover, their own location, the time and brightness of each meteor, and whether it was from the main meteor stream. It is also worth noting details of persistent afterglows (trains) and fireballs, and making counts of how many meteors appear in a given period.

COMETS

Comets are small bodies in orbit about the Sun. Consisting of frozen gases and dust, they are often known as 'dirty snowballs'. When their orbits bring them close to the Sun, the ices evaporate and dramatic tails of gas and dust can sometimes be seen.

A number of comets move round the Sun in fairly small, elliptical orbits in periods of a few years; others have much longer periods. Most really brilliant comets have orbital periods of several thousands or even millions of years. The exception is Comet Halley, a bright comet with a period of about 76 years. It was last seen with the naked eye in 1986.

Binoculars and wide-field telescopes provide the best views of comet tails. Larger telescopes with a high magnification are necessary to observe fine detail in the gaseous head (*coma*). Most comets are discovered with professional instruments, but a few are still found by experienced amateur astronomers.

None of the known comets is predicted to reach naked-eye brightness in 2011, but there's always a chance of a bright new comet putting in a surprise appearance.

Deep-sky objects are 'fuzzy patches' that lie outside the Solar System. They include star clusters, nebulae and galaxies. To observe the majority of deep-sky objects you will need binoculars or a telescope, but there are also some beautiful naked-eye objects, notably the Pleiades and the Orion Nebula.

The faintest object that an instrument can see is its *limiting magnitude*. The table gives a rough guide, for good seeing conditions, for a variety of small- to medium-sized telescopes.

We have provided a selection of recommended deep-sky targets, together with their magnitudes. Some are described in more detail in our monthly 'Object' features. Look on the appropriate month's map to find which constellations are on view, and then choose your objects using the list below. We have provided celestial coordinates for readers with detailed star maps. The suggested times of year for viewing are when the constellation is highest in the sky in the late evening.

Limiting magnitude for small to medium telescopes	
Aperture (mm)	Limiting magnitude
50	+11.2
60	+11.6
70	+11.9
80	+12.2
100	+12.7
125	+13.2
150	+13.6

RECOMMENDED DEEP-SKY OBJECTS

Andromeda – autumn and early winter

M31 (NGC 224) Andromeda Galaxy	3rd-magnitude spiral galaxy RA 00h 42.7m Dec +41° 16'
M32 (NGC 221)	8th-magnitude elliptical galaxy, a companion to M31 RA 00h 42.7m Dec +40° 52'
M110 (NGC 205)	8th-magnitude elliptical galaxy RA 00h 40.4m Dec +41° 41'
NGC 7662 Blue Snowball	8th-magnitude planetary nebula RA 23h 25.9m Dec +42° 33'

Aquarius – late autumn and early winter

M2 (NGC 7089)	6th-magnitude globular cluster RA 21h 33.5m Dec –00° 49'
M72 (NGC 6981)	9th-magnitude globular cluster RA 20h 53.5m Dec –12° 32'
NGC 7293 Helix Nebula	7th-magnitude planetary nebula RA 22h 29.6m Dec –20° 48'
NGC 7009 Saturn Nebula	8th-magnitude planetary nebula RA 21h 04.2m Dec –11° 22'

Aries – early winter

NGC 772	10th-magnitude spiral galaxy RA 01h 59.3m Dec +19° 01'

Auriga – winter

M36 (NGC 1960)	6th-magnitude open cluster RA 05h 36.1m Dec +34° 08'
M37 (NGC 2099)	6th-magnitude open cluster RA 05h 52.4m Dec +32° 33'
M38 (NGC 1912)	6th-magnitude open cluster RA 05h 28.7m Dec +35° 50'

Cancer – late winter to early spring

M44 (NGC 2632) Praesepe or Beehive	3rd-magnitude open cluster RA 08h 40.1m Dec +19° 59'
M67 (NGC 2682)	7th-magnitude open cluster RA 08h 50.4m Dec +11° 49'

Canes Venatici – visible all year

M3 (NGC 5272)	6th-magnitude globular cluster RA 13h 42.2m Dec +28° 23'

M51 (NGC 5194/5) Whirlpool Galaxy	8th-magnitude spiral galaxy RA 13h 29.9m Dec +47° 12'
M63 (NGC 5055)	9th-magnitude spiral galaxy RA 13h 15.8m Dec +42° 02'
M94 (NGC 4736)	8th-magnitude spiral galaxy RA 12h 50.9m Dec +41° 07'
M106 (NGC4258)	8th-magnitude spiral galaxy RA 12h 19.0m Dec +47° 18'

Canis Major – late winter

M41 (NGC 2287)	4th-magnitude open cluster RA 06h 47.0m Dec –20° 44'

Capricornus – late summer and early autumn

M30 (NGC 7099)	7th-magnitude globular cluster RA 21h 40.4m Dec –23° 11'

Cassiopeia – visible all year

M52 (NGC 7654)	6th-magnitude open cluster RA 23h 24.2m Dec +61° 35'
M103 (NGC 581)	7th-magnitude open cluster RA 01h 33.2m Dec +60° 42'
NGC 225	7th-magnitude open cluster RA 00h 43.4m Dec +61 47'
NGC 457	6th-magnitude open cluster RA 01h 19.1m Dec +58° 20'
NGC 663	Good binocular open cluster RA 01h 46.0m Dec +61° 15'

Cepheus – visible all year

Delta Cephei	Variable star, varying between +3.5 and +4.4 with a period of 5.37 days. It has a magnitude +6.3 companion and they make an attractive pair for small telescopes or binoculars.

Cetus – late autumn

Mira (omicron Ceti)	Irregular variable star with a period of roughly 330 days and a range between +2.0 and +10.1.
M77 (NGC 1068)	9th-magnitude spiral galaxy RA 02h 42.7m Dec –00° 01'

Coma Berenices – spring

M53 (NGC 5024)	8th-magnitude globular cluster RA 13h 12.9m Dec +18° 10'
M64 (NGC 4286) Black Eye Galaxy	8th-magnitude spiral galaxy with a prominent dust lane that is visible in larger telescopes. RA 12h 56.7m Dec +21° 41'
M85 (NGC 4382)	9th-magnitude elliptical galaxy RA 12h 25.4m Dec +18° 11'
M88 (NGC 4501)	10th-magnitude spiral galaxy RA 12h 32.0m Dec.+14° 25'
M91 (NGC 4548)	10th-magnitude spiral galaxy RA 12h 35.4m Dec +14° 30'
M98 (NGC 4192)	10th-magnitude spiral galaxy RA 12h 13.8m Dec +14° 54'
M99 (NGC 4254)	10th-magnitude spiral galaxy RA 12h 18.8m Dec +14° 25'
M100 (NGC 4321)	9th-magnitude spiral galaxy RA 12h 22.9m Dec +15° 49'
NGC 4565	10th-magnitude spiral galaxy RA 12h 36.3m Dec +25° 59'

Cygnus – late summer and autumn

Cygnus Rift	Dark cloud just south of Deneb that appears to split the Milky Way in two.
NGC 7000 North America Nebula	A bright nebula against the background of the Milky Way, visible with binoculars under dark skies. RA 20h 58.8m Dec +44° 20'
NGC 6992 Veil Nebula (part)	Supernova remnant, visible with binoculars under dark skies. RA 20h 56.8m Dec +31 28'
M29 (NGC 6913)	7th-magnitude open cluster RA 20h 23.9m Dec +36° 32'
M39 (NGC 7092)	Large 5th-magnitude open cluster RA 21h 32.2m Dec +48° 26'
NGC 6826 Blinking Planetary	9th-magnitude planetary nebula RA 19 44.8m Dec +50° 31'

Delphinus – late summer

NGC 6934	9th-magnitude globular cluster RA 20h 34.2m Dec +07° 24'

Draco – midsummer

NGC 6543	9th-magnitude planetary nebula RA 17h 58.6m Dec +66° 38'

Gemini – winter

M35 (NGC 2168)	5th-magnitude open cluster RA 06h 08.9m Dec +24° 20'
NGC 2392 Eskimo Nebula	8–10th-magnitude planetary nebula RA 07h 29.2m Dec +20° 55'

Hercules – early summer

M13 (NGC 6205)	6th-magnitude globular cluster RA 16h 41.7m Dec +36° 28'
M92 (NGC 6341)	6th-magnitude globular cluster RA 17h 17.1m Dec +43° 08'
NGC 6210	9th-magnitude planetary nebula RA 16h 44.5m Dec +23 49'

Hydra – early spring

M48 (NGC 2548)	6th-magnitude open cluster RA 08h 13.8m Dec −05° 48'
M68 (NGC 4590)	8th-magnitude globular cluster RA 12h 39.5m Dec −26° 45'

M83 (NGC 5236)	8th-magnitude spiral galaxy RA 13h 37.0m Dec −29° 52'
NGC 3242 Ghost of Jupiter	9th-magnitude planetary nebula RA 10h 24.8m Dec −18°38'

Leo – spring

M65 (NGC 3623)	9th-magnitude spiral galaxy RA 11h 18.9m Dec +13° 05'
M66 (NGC 3627)	9th-magnitude spiral galaxy RA 11h 20.2m Dec +12° 59'
M95 (NGC 3351)	10th-magnitude spiral galaxy RA 10h 44.0m Dec +11° 42'
M96 (NGC 3368)	9th-magnitude spiral galaxy RA 10h 46.8m Dec +11° 49'
M105 (NGC 3379)	9th-magnitude elliptical galaxy RA 10h 47.8m Dec +12° 35'

Lepus – winter

M79 (NGC 1904)	8th-magnitude globular cluster RA 05h 24.5m Dec −24° 33'

Lyra – spring

M56 (NGC 6779)	8th-magnitude globular cluster RA 19h 16.6m Dec +30° 11'
M57 (NGC 6720) Ring Nebula	9th-magnitude planetary nebula RA 18h 53.6m Dec +33° 02'

Monoceros – winter

M50 (NGC 2323)	6th-magnitude open cluster RA 07h 03.2m Dec −08° 20'
NGC 2244	Open cluster surrounded by the faint Rosette Nebula, NGC 2237. Visible in binoculars. RA 06h 32.4m Dec +04° 52'

Ophiuchus – summer

M9 (NGC 6333)	8th-magnitude globular cluster RA 17h 19.2m Dec −18° 31'
M10 (NGC 6254)	7th-magnitude globular cluster RA 16h 57.1m Dec −04° 06'
M12 (NCG 6218)	7th-magnitude globular cluster RA 16h 47.2m Dec −01° 57'
M14 (NGC 6402)	8th-magnitude globular cluster RA 17h 37.6m Dec −03° 15'
M19 (NGC 6273)	7th-magnitude globular cluster RA 17h 02.6m Dec −26° 16'
M62 (NGC 6266)	7th-magnitude globular cluster RA 17h 01.2m Dec −30° 07'
M107 (NGC 6171)	8th-magnitude globular cluster RA 16h 32.5m Dec −13° 03'

Orion – winter

M42 (NGC 1976) Orion Nebula	4th-magnitude nebula RA 05h 35.4m Dec −05° 27'
M43 (NGC 1982)	5th-magnitude nebula RA 05h 35.6m Dec −05° 16'
M78 (NGC 2068)	8th-magnitude nebula RA 05h 46.7m Dec +00° 03'

Pegasus – autumn

M15 (NGC 7078)	6th-magnitude globular cluster RA 21h 30.0m Dec +12° 10'

Perseus – autumn to winter

M34 (NGC 1039)	5th-magnitude open cluster RA 02h 42.0m Dec +42° 47'
M76 (NGC 650/1) Little Dumbbell	11th-magnitude planetary nebula RA 01h 42.4m Dec +51° 34'

NGC 869/884
Double Cluster
Pair of open star clusters
RA 02h 19.0m Dec +57° 09'
RA 02h 22.4m Dec +57° 07'

Pisces – autumn

M74 (NGC 628)
9th-magnitude spiral galaxy
RA 01h 36.7m Dec +15° 47'

Puppis – late winter

M46 (NGC 2437)
6th-magnitude open cluster
RA 07h 41.8m Dec –14° 49'

M47 (NGC 2422)
4th-magnitude open cluster
RA 07h 36.6m Dec –14° 30'

M93 (NGC 2447)
6th-magnitude open cluster
RA 07h 44.6m Dec –23° 52'

Sagitta – late summer

M71 (NGC 6838)
8th-magnitude globular cluster
RA 19h 53.8m Dec +18° 47'

Sagittarius – summer

M8 (NGC 6523)
Lagoon Nebula
6th-magnitude nebula
RA 18h 03.8m Dec –24° 23'

M17 (NGC 6618)
Omega Nebula
6th-magnitude nebula
RA 18h 20.8m Dec –16° 11'

M18 (NGC 6613)
7th-magnitude open cluster
RA 18h 19.9m Dec –17 08'

M20 (NGC 6514)
Trifid Nebula
9th-magnitude nebula
RA 18h 02.3m Dec –23° 02'

M21 (NGC 6531)
6th-magnitude open cluster
RA 18h 04.6m Dec –22° 30'

M22 (NGC 6656)
5th-magnitude globular cluster
RA 18h 36.4m Dec –23° 54'

M23 (NGC 6494)
5th-magnitude open cluster
RA 17h 56.8m Dec –19° 01'

M24 (NGC 6603)
5th-magnitude open cluster
RA 18h 16.9m Dec –18° 29'

M25 (IC 4725)
5th-magnitude open cluster
RA 18h 31.6m Dec –19° 15'

M28 (NGC 6626)
7th-magnitude globular cluster
RA 18h 24.5m Dec –24° 52'

M54 (NGC 6715)
8th-magnitude globular cluster
RA 18h 55.1m Dec –30° 29'

M55 (NGC 6809)
7th-magnitude globular cluster
RA 19h 40.0m Dec –30° 58'

M69 (NGC 6637)
8th-magnitude globular cluster
RA 18h 31.4m Dec –32° 21'

M70 (NGC 6681)
8th-magnitude globular cluster
RA 18h 43.2m Dec –32° 18'

M75 (NGC 6864)
9th-magnitude globular cluster
RA 20h 06.1m Dec –21° 55'

Scorpius (northern part) – midsummer

M4 (NGC 6121)
6th-magnitude globular cluster
RA 16h 23.6m Dec –26° 32'

M7 (NGC 6475)
3rd-magnitude open cluster
RA 17h 53.9m Dec –34° 49'

M80 (NGC 6093)
7th-magnitude globular cluster
RA 16h 17.0m Dec –22° 59'

Scutum – mid to late summer

M11 (NGC 6705)
Wild Duck Cluster
6th-magnitude open cluster
RA 18h 51.1m Dec –06° 16'

M26 (NGC 6694)
8th-magnitude open cluster
RA 18h 45.2m Dec –09° 24'

Serpens – summer

M5 (NGC 5904)
6th-magnitude globular cluster
RA 15h 18.6m Dec +02° 05'

M16 (NGC 6611)
6th-magnitude open cluster,
surrounded by the Eagle Nebula.
RA 18h 18.8m Dec –13° 47'

Taurus – winter

M1 (NGC 1952)
Crab Nebula
8th-magnitude supernova remnant
RA 05h 34.5m Dec +22° 00'

M45
Pleiades
1st-magnitude open cluster,
an excellent binocular object.
RA 03h 47.0m Dec +24° 07'

Triangulum – autumn

M33 (NGC 598)
6th-magnitude spiral galaxy
RA 01h 33.9m Dec +30° 39'

Ursa Major – all year

M81 (NGC 3031)
7th-magnitude spiral galaxy
RA 09h 55.6m Dec +69° 04'

M82 (NGC 3034)
8th-magnitude starburst galaxy
RA 09h 55.8m Dec +69° 41'

M97 (NGC 3587)
Owl Nebula
12th-magnitude planetary nebula
RA 11h 14.8m Dec +55° 01'

M101 (NGC 5457)
8th-magnitude spiral galaxy
RA 14h 03.2m Dec +54° 21'

M108 (NGC 3556)
10th-magnitude spiral galaxy
RA 11h 11.5m Dec +55° 40'

M109 (NGC 3992)
10th-magnitude spiral galaxy
RA 11h 57.6m Dec +53° 23'

Virgo – spring

M49 (NGC 4472)
8th-magnitude elliptical galaxy
RA 12h 29.8m Dec +08° 00'

M58 (NGC 4579)
10th-magnitude spiral galaxy
RA 12h 37.7m Dec +11° 49'

M59 (NGC 4621)
10th-magnitude elliptical galaxy
RA 12h 42.0m Dec +11° 39'

M60 (NGC 4649)
9th-magnitude elliptical galaxy
RA 12h 43.7m Dec +11° 33'

M61 (NGC 4303)
10th-magnitude spiral galaxy
RA 12h 21.9m Dec +04° 28'

M84 (NGC 4374)
9th-magnitude elliptical galaxy
RA 12h 25.1m Dec +12° 53'

M86 (NGC 4406)
9th-magnitude elliptical galaxy
RA 12h 26.2m Dec +12° 57'

M87 (NGC 4486)
9th-magnitude elliptical galaxy
RA 12h 30.8m Dec +12° 24'

M89 (NGC 4552)
10th-magnitude elliptical galaxy
RA 12h 35.7m Dec +12° 33'

M90 (NGC 4569)
9th-magnitude spiral galaxy
RA 12h 36.8m Dec +13° 10'

M104 (NGC 4594)
Sombrero Galaxy
Almost edge-on 8th-magnitude
spiral galaxy.
RA 12h 40.0m Dec –11° 37'

Vulpecula – late summer and autumn

M27 (NGC 6853)
Dumbbell Nebula
8th-magnitude planetary nebula
RA 19h 59.6m Dec +22° 43'

When choosing a telescope, it's always a good idea to set out your requirements first. How large a telescope can you cope with? What are your local conditions like? What are your observational interests? What's your budget? To this list, people often add that they would like to be able to take photos through their telescope. But this seemingly simple requirement can make a huge difference to your choice, and can easily end up being the dominant factor. So this article outlines the basics of what you need for astrophotography of different types, from the point of view of choosing the right kit.

Actually, astrophotography of all kinds has never been easier. Today, even suburban amateurs with fairly modest means can take pictures as good as or better than those which once came from the leading observatories. All this comes at a cost in cameras, equipment and time. So it's important to be aware before you start how long is the road which you have decided to take! And maybe there are ways of getting quite impressive results without going the whole hog.

Photographing the Moon

The Moon is by far the easiest object to get started with. It's bright, easy to find in any telescope and often yields spectacular results. At its simplest, you can aim your scope at the Moon, point your camera into the eyepiece and take a snap. Quite often the result is impressive, and even mobile phone cameras can give stunning shots if you're lucky.

A more advanced approach is to use a single-lens reflex (SLR) camera, which allows you to remove its normal lens and attach it directly to the telescope in place of the eyepiece. The telescope then behaves exactly the same way as a long telephoto lens. You can get adapters for most modern digital SLR (DSLR) cameras, usually on the T-mount or T-ring system.

Once your camera is attached, you can snap away, ideally using a cable release so as not to jog the telescope when pressing the shutter button. In the case of the Moon, a typical exposure time will be a fraction of a second at ISO 100, during which time the Moon will not have drifted significantly through the field of view. So you don't need any kind of drive on the telescope.

Photographing the planets

The set-up described above will work equally well with the bright planets. But most planetary photographers don't actually use weighty SLR cameras for their work. Instead, they use purpose-

▼ *A DSLR camera attached directly to the telescope using an adapter. Notice that the camera focus point is fairly close to the telescope, and some focusers do not have enough travel to allow this.*

made cameras based on webcam technology. Webcams are basically small video cameras which are used for chatting online. They provide a compressed video stream at much lower resolution than a full-sized camera, but they are lightweight and can give very good images.

Off-the-shelf webcams require a lot of modification before they can be attached to a telescope, but Celestron have their own purpose-built planetary camera, the NexImage, which is sold complete with an eyepiece adapter and software for linking it to the computer which is essential when undertaking this type of photography. Most people use a laptop computer for the purpose, usually near to the telescope. More advanced cameras are available which don't compress the individual images when producing the datastream, though at somewhat higher cost. The NexImage costs about £140 and more advanced models from The Imaging Source may cost two or three times that amount.

A webcam has a much smaller field of view than a DSLR camera, though the detail shown might be similar. But where the webcam scores is in its ability to record many frames per second in real time. As you watch the planet on the monitor, you can see that the image is usually both moving around slightly and becoming distorted as a result of the turbulence in our atmosphere – what astronomers call 'seeing'. With a webcam it's easy to record not just individual images but hundreds in a short time. Freely available software, such as Registax or AVIStack, then selects the best images and stacks them so as to reduce the electronic noise in the images. Image processing then sharpens the image, giving results of excellent quality even from small instruments.

Ideally, your telescope should have a motor drive for planetary photography, so as to keep the planet on the webcam's tiny chip for a few minutes at a time. But a precise drive isn't essential, as the software is designed to cope with some image movement.

Photographing faint objects

Most deep-sky objects, and many Solar System objects such as comets, are much fainter than the Moon and planets, and require exposure times of the order of many seconds to minutes. Webcams are usually limited to brief exposures, but DSLR cameras can usually give exposures as long as you like. You can also get purpose-built

SUPPLIERS OF THE EQUIPMENT MENTIONED HERE

Sky-Watcher telescopes and autoguiders from Optical Vision: check www.opticalvision.co.uk for dealers

Meade telescopes and camera adapters from Telescope House: www.telescopehouse.com

Celestron NexImage: check www.celestron.uk.com for dealers

Autoguiding CCDs and accessories from Opticstar: www.opticstar.com

Prices quoted here are current as of mid-2010.

▼ A webcam is lightweight and can be attached to virtually any telescope. More advanced cameras give better images, and the best results come from mono cameras that require images to be taken through separate colour filters.

▲ *Peter Shah used this 200 mm reflecting telescope to take the images on pages 8, 40 and 53. The CCD camera is attached to the main scope, while the smaller refracting telescope has an autoguider (not visible).*

CCD cameras with cooled chips to reduce electronic noise, and these are capable of giving really superb results, though they cost far more than a popular DSLR – in some cases thousands of pounds.

Apart from a camera, there are two prime requirements when taking long-exposure photos. One is an equatorial mount, which is aligned with the Earth's axis so as to counteract the daily turning of the heavens. A simple altazimuth mount, with up-down and left-right axes, as used on many instruments, is of little use for long exposures because the field of view will rotate along with the sky. The other requirement is an accurate motor drive. But the telescope drive that keeps Jupiter within the field of view for planetary imaging will probably be hopeless for exposure times longer than a few tens of seconds. Even a tiny error will be enough to turn a star image from a point into a streak.

Until recently, long exposures involved heroic efforts manually guiding on a star to overcome such errors. A few seconds' inattention at the end of a one-hour exposure could ruin the whole thing! But these days, autoguiding is the solution. This generally means having a separate telescope on the same mounting with a purpose-built autoguider attached, though there are ways in which you can use the same telescope. An autoguider consists of a separate CCD unit or even a webcam which takes an image of a star every second or so. Any drift of the star's image causes it to correct the drive rate to counteract it. Generally, you need a separate program running on your laptop to accomplish the autoguiding, but stand-alone autoguiders are now coming on to the market. These have a small display which shows the locations of stars in the field of view, from which you select one.

Autoguiders will work with a variety of the more advanced mounts which have separate autoguiding ports. These are often referred to as ST-4 ports, after the SBIG ST-4 unit which was the first widely used autoguider. Typical mounts include the Sky-Watcher Go To mounts such as the HEQ5, and the Celestron CG-5GT. Meade LX200 Schmidt-Cassegrain instruments are already on fork mounts which also have suitable autoguiding ports, and there are Meade accessories which will attach a separate telescope to the mounts. But if you are using an equatorial mount, you'll need to go to a third-party supplier for a suitable bar allowing you to put two telescopes on to the same mount. Many users also swear by the free PHD autoguiding software (www.stark-labs.com) which has clever autocalibration, rather than the software that comes with their autoguider.

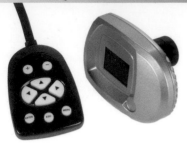

▲ Sky-Watcher's new Synguider, introduced in 2010, is stand-alone and needs no laptop. Once in focus, the screen at the back shows the positions of the stars in the field of view.

But the whole business starts to get horribly complicated and indeed expensive. You need a suitable mount, your main telescope, a guide telescope, an autoguider, a camera to take the images and a laptop to control the whole thing, which together will probably cost at least £5,000. You have to polar-align your mount, connect up everything including power supplies for the mount, laptop and cameras, and then hope that it is still clear by the time you get round to taking the exposures! As one expert warned me, 'Don't forget the medical bills for stress as well!' But the upside is that, with perseverance, you can take superb images.

However, there is an easier way out, and that is to take images that are within the limitations of your equipment. The problems all stem from trying to take pictures at longer and longer focal lengths, which then require precision guiding. But if you use shorter focal lengths – say an ordinary 200 mm telephoto lens instead of your telescope – then you can get away with much less effort, and still take some striking photos. All you need to do is to piggyback your camera on top of the telescope. The drive will probably allow exposures of several minutes without any guiding. The field of view of a 200 mm lens on the average DLSR is just under 7 × 5°, which is about the same as many binoculars. Many star clusters, nebulae and even galaxies are well within its grasp, and such a combination is ideal if a good comet comes along. After all, the main thing is to get results which look good and might be worth framing for your wall!

▲ Telephoto lenses are ideal for wide-field shots, such as this 2005 view of the Pleiades star cluster and Comet Machholz, taken by Thierry Legault with a Canon 10D camera and 200 mm lens. The exposure time was 5 minutes at ISO 400.

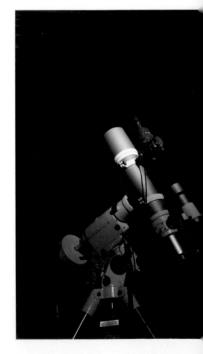

► Many telescope mounts include a threaded rod on the tube rings which allow you to attach a camera with a telephoto lens. Driven exposures of many minutes are often possible without autoguiding.